Permeable Barriers for Groundwater Remediation

Design, Construction, and Monitoring

by

Arun R. Gavaskar
Neeraj Gupta
Bruce M. Sass
Robert J. Janosy
Dennis O'Sullivan

BATTELLE PRESS

Columbus • Richland

Library of Congress Cataloging-in-Publication Data

Gavaskar, Arun R., 1962–
    Permeable barriers for groundwater remediation / by Arun R. Gavaskar
    ... [et al.].
        p.   cm.
    Includes bibliographical references and index.
    ISBN 1-57477-036-5 (alk. paper)
    1. Groundwater –Purification. 2. Membranes (Technology)
    I. Title.
    TD426.G38   1998
    628.1'68—dc21   97-20764
                                                                                                      CIP

For additional Battelle books and software on remediation, see "Books from Battelle Press" at www.battelle.org/bookstore.

    Battelle Press
    505 King Avenue
    Columbus, Ohio 43201, USA
    1-800-451-3543 or 614-424-6393
    Fax: 614-424-3819
    e-mail: press@battelle.org

ACKNOWLEDGMENTS

The contents of this book were prepared by Battelle Memorial Institute, 505 King Avenue, Columbus, Ohio 43201, under Contract No. F08697-95-D-6004 (DO 5501) for the Armstrong Laboratory Environics Directorate (AL/EQ), 139 Barnes Drive, Tyndall Air Force Base, Florida 32403.

This book provides design guidance for construction and monitoring of a permeable barrier for remediation of contaminated groundwater. An objective of this book is to bring together the existing knowledge base (published and unpublished) on this technology. This is intended to be a stand-alone document that provides guidance to site managers, contractors, regulators, and academics. Full citations for the references are included in Chapter 10.

Battelle would like to acknowledge the advice and support provided by representatives of the Air Force, including Captain Jeff Stinson (who served as the Project Officer during part of the preparation of this document), Dave Burris, Captain Edward Marchand, and Randy Wolf (BDM, Inc.). Also providing support were Steve White and Ian Osgerby from the U.S. Army Corps of Engineers. We appreciate the cooperation and advice provided by several Remediation Technologies Development Forum (RTDF) members, especially John Vogan, Stephanie O'Hannesin, and Robert Focht of Envirometal Technologies, Inc. (ETI), Scott Warner of Geomatrix, Tim Sivavec of General Electric Company, and Charles Reeter of Naval Facilities Engineering Service Center (NFESC). Also providing significant informational support were Pat Mackenzie of General Electric Company, Dale Schultz and Rich Landis of DuPont, Steve McCutcheon of the U.S. Environmental Protection Agency (EPA) National Exposure Research Laboratory, and Robert Puls of EPA National Risk Management Research Laboratory. Several members of the Interstate Technology and Regulatory Cooperation (ITRC) Permeable Barriers Subgroup, including Matt Turner of the New Jersey Department of Environmental Protection, Robert Hanson of Coleman Research Corporation, David LaPusata and Anna Symington of the Massachusetts Department of Environmental Protection, Paul Hadley of the California Department of Toxic Substances Control, and Brad Job and Hatbe Kilfe of the California Regional Water Quality Control Board provided important information and review. We appreciate their support and cooperation.

Other Battelle staff who reviewed this book in draft and contributed to its contents include Tad Fox, Kirk Cantrell, James Hicks, and Robert Olfenbuttel. Tim Lundgren provided editing support, Loretta Bahn provided desktop publishing support, and Vic Saylor provided graphics support.

CONTENTS

Appendices

Figures

Tables

ABBREVIATIONS, ACRONYMS, AND SYMBOLS

2D	two-dimensional
3D	three-dimensional
AFB	Air Force Base
AL/EQ	Armstrong Laboratory Environics Directorate
BET	Brunauer-Emmett-Teller adsorption isotherm equation
bgs	below ground surface
BTEX	benzene, toluene, ethylbenzene, and xylenes
cDCE	cis-1,2-dichloroethene
CMC	carboxymethyl cellulose
CMS	Corrective Measures Study
CP	cone penetrometer
CPT	cone penetrometer test
CQC	construction quality control
DCE	dichloroethene
DNAPL	dense, nonaqueous-phase liquid
DO	dissolved oxygen
DOC	dissolved organic carbon
DoD	Department of Defense
DOE	Department of Energy
ECD	electron capture detector
EDS	energy dispersive x-ray spectroscopy
Eh	redox potential
EPA	Environmental Protection Agency
EQL	estimated quantitation limit
ETI	Envirometal Technologies, Inc.
FGDM	Funnel-and-Gate Design Model
GC	gas chromatography
GC-FID	gas chromatograph-flame ionization detector
GE	General Electric
GX	gum xanthan
HDPE	high-density polyethylene
HFB	horizontal flow barrier
HSU	hydrostratigraphic units

IAP	ion activity product
IC	ion chromatography
ICP	inductively coupled plasma
ITRC	Interstate Technology and Regulatory Cooperation
K	hydraulic conductivity
$K_{aquifer}$	aquifer hydraulic conductivity
K_{cell}	reactive cell hydraulic conductivity
MCL	maximum contaminant level
MFA	Moffett Federal Airfield
MS	matrix spike
MSDS	Material Safety Data Sheet
NA	not applicable
NERL	National Exposure Research Laboratory
NFESC	Naval Facilities Engineering Service Center
NPV	net present value
NRC	National Research Council
O&M	operating and maintenance
ORC	oxygen-releasing compound
OSHA	Occupational Safety and Health Administration
PBWG	Permeable Barriers Working Group
PCB	polychlorinated biphenyl
PCE	perchloroethylene
PPE	personal protective equipment
PRP	potentially responsible party
PVC	polyvinyl chloride
QA	quality assurance
QAPP	Quality Assurance Project Plan
QC	quality control
RACER	Remedial Action Cost Engineering and Requirements
RCRA	Resource Conservation and Recovery Act
RFI	RCRA Facility Investigation
RFI/CMS	RFI/Corrective Measures Study
RI/FS	Remedial Investigation/Feasibility Study
ROD	Record of Decision
RTDF	Remediation Technologies Development Forum
SEM	scanning electron microscopy
SI	saturation index

SITE	Superfund Innovative Technology Evaluation
SPH	smooth particle hydrodynamics
T	temperature
$t_{1/2}$	half-life
TCA	trichloroethane
TCE	trichloroethylene
t-DCE	*trans*-1,2-dichloroethene
TDS	total dissolved solids
TOC	total organic carbon
TSS	total suspended solids
USCG	United States Coast Guard
U.S. DOE	U.S. Department of Energy
U.S. EPA	U.S. Environmental Protection Agency
USGS	United States Geological Survey
UST	underground storage tank
VC	vinyl chloride
VOC	volatile organic compound
VP	vinyl polymer
WDS	wave dispersive spectroscopy
XRD	x-ray diffraction

Permeable Barriers for Groundwater Remediation

Design, Construction, and Monitoring

CHAPTER 1

TECHNOLOGY BACKGROUND AND STATUS

1.1 PROBLEM DESCRIPTION

Chlorinated solvents have been used extensively in the past by industry and government for a variety of operations, such as degreasing, maintenance, and dry cleaning. Leaks and spills of these solvents, as well as historical disposal practices, have led to widespread contamination of the soil and groundwater. Ten of the twenty-five most common groundwater contaminants at hazardous waste sites are chlorinated solvents, with trichloroethylene (TCE) being the most prevalent (National Research Council, 1994).

Most chlorinated solvents belong to a class of compounds which, when present in sufficient quantity, may form dense, nonaqueous-phase liquids (DNAPLs). DNAPLs are denser than water and therefore move downward in the subsurface until they encounter a low-permeability zone or aquitard. On their way down, DNAPLs typically leave a trail of free-phase residual DNAPL that is virtually immobile and resistant to pumping. The DNAPLs present in pools or in residually saturated zones provide a long-term source for contaminant release into the groundwater, often resulting in large dissolved-phase plumes. Although most chlorinated solvents are sparingly soluble in water, their solubilities are several times higher than the United States Environmental Protection Agency's (U.S. EPA's) maximum contaminant level (MCL) standards for drinking water. Table 1-1 shows the properties of common chlorinated solvents. Because of their low solubilities and mass transfer limitations, chlorinated solvent source zones persist in the aquifer for several years, decades, or even centuries. The dissolved contaminant plume resulting from the source zone persists for similar lengths of time and has been known to travel large distances, because chlorinated solvents are relatively resistant to biodegradation processes at many sites.

Although one apparent approach would be to remediate the DNAPL source zone, in practice this often proves difficult. First, DNAPL source zones are difficult to find; when found, they are generally difficult to remediate. Therefore, at many sites, a more viable option is to treat the plume. Conventional pump-and-treat systems could be used to capture and treat the plume. However, past experience at contaminated groundwater sites and recent studies (National Research Council, 1994) have shown the inadequacies of this approach. Also, pump-and-treat systems would have to be operated for many years or decades, or as long as the source zone and plume persist. The associated operational costs over several

TABLE 1-1. Properties of common chlorinated organic compounds

Compound	MCL (mg/L)	Water Solubility (mg/L at 25°C)	Density (g/cm³ at 20°C)	Vapor Pressure (Pascals at 25°C)
Carbon tetrachloride	0.005	800	1.59	15,097
1,2-dichloroethane	0.005	8,600	1.26	9,000
Methylene chloride	0.005	20,000	1.33	46,522 (20°C)
Perchloroethylene	0.005	150	1.63	2,415
1,1,1-trichloroethane	0.2	1,250	1.34	13,300
Trichlorethylene	0.005	1,100	1.46	9,910
cis-1,2-dichloroethene	0.07	3,500	1.28	26,700
trans-1,2-dichloroethene	0.1	6,300	1.26	45,300
Vinyl chloride	0.002	2,000	0.91	350,000

decades can be enormous. The recent development of permeable reactive barriers has presented a potentially viable alternative to conventional pump-and-treat systems for remediation of chlorinated solvent-contaminated groundwater. Additionally, dissolved metals, such as chromium, and petroleum hydrocarbons are being targeted for treatment with permeable barriers.

This book provides guidance for the design, construction, and monitoring of a permeable barrier for remediation of contaminated groundwater. In addition, it serves as a compilation of the existing knowledge base (published and unpublished) regarding the use of this technology for groundwater remediation. This stand-alone document is intended to provide guidance to site managers, contractors, regulators, and academics in regard to both the application of permeable barriers and research.

1.2 TECHNOLOGY DESCRIPTION

Figure 1-1 shows some possible configurations of permeable barrier systems. In its simplest form, a permeable reactive barrier consists of a zone of reactive material, such as granular iron, installed in the path of a dissolved chlorinated solvent plume (Figure 1-1a). As the groundwater flows through this permeable barrier, the chlorinated organics come in contact with the reactive medium and are degraded to potentially nontoxic dehalogenated organic compounds and inorganic chloride. The main advantage of this system is that, generally, no pumping or aboveground treatment is required; the barrier acts passively after installation. Because there are no aboveground installed structures, the affected property can be put to productive use while it is being cleaned up. Initial evidence indicates that the reactive medium is used up very slowly and, therefore, permeable reactive barriers have the potential to passively treat the plume over several years or decades. This would result in hardly any annual operating costs, other than site

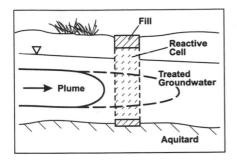

(a) Elevation view of a permeable barrier

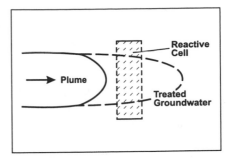

(b) Plan view of a continuous reactive barrier configuration

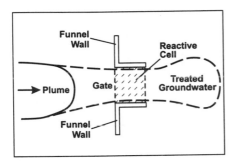

(c) Funnel-and-gate system (plan view)

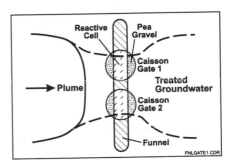

(d) Funnel-and-gate system with two caisson gates (plan view)

FIGURE 1-1. Schematic illustrations of some permeable barrier configurations

monitoring. Depending on the longevity of the reactive medium, the barrier may have to be rejuvenated or replaced periodically; however, it is expected that this maintenance would be required relatively infrequently.

A permeable barrier may be installed as a continuous reactive barrier or as a funnel-and-gate system. A continuous reactive barrier (Figure 1-1b) consists of a reactive cell containing the permeable reactive medium. A funnel-and-gate system (Figure 1-1c) has an impermeable section (or funnel) that directs the captured groundwater flow towards the permeable section (or gate). This configuration sometimes allows better control over reactive cell placement and plume capture. At sites where the groundwater flow is very heterogeneous, a funnel-and-gate system can allow the reactive cell to be placed in the more permeable portions of the aquifer. At sites where the contaminant distribution is very nonuniform, a funnel-and-gate system can better homogenize the concentrations of contaminants entering the reactive cell.

A system with multiple gates (Figure 1-1d) can be used to ensure sufficient residence time at sites with a relatively wide plume and high groundwater velocity, especially when the size of each reactive cell or gate is limited by the method of emplacement (e.g., emplacement with caissons). Hydraulic modeling conducted by Starr and Cherry (1994) has shown that the most efficient funnel-and-gate arrangement is with the funnel walls aligned in a straight line with the gate (Figure 1-2a). However, other funnel-and-gate arrangements (Figure 1-2b) are possible and have been employed during previous applications. Property boundaries and geotechnical considerations (e.g., presence of underground utilities) may sometimes govern the shape of the funnel.

1.3 MECHANISM OF ENHANCED ABIOTIC DEGRADATION WITH METALS

Although a variety of reactive media (see Chapter 3) could be used to treat groundwater contaminants, the most commonly used medium so far has been zero-valent granular iron. As the zero-valent metal in the reactive cell corrodes, the resulting electron activity is believed to reduce the chlorinated compounds to potentially nontoxic products. Because granular iron is the only reactive medium that has been used so far in field applications and because the mechanism of chlorinated solvent degradation with zero-valent iron has been the most widely studied and reported to date, most of the discussion on abiotic destruction in this book focuses on the use of granular iron as the reactive medium. Other zero-valent metals may exhibit similar reactions with differing rates.

The first reported use of the degradation potential of metals for treating chlorinated organic compounds in the environment was by Sweeny and Fischer (1972), who acquired a patent for the degradation of chlorinated pesticides by metallic zinc under acidic conditions. These researchers found that p,p'-DDT was degraded by zinc at ambient temperatures at a satisfactory rate with ethane as the major product. In two later papers, Sweeny (1981a and b) described how catalytically active powders of iron, zinc. or aluminum could be used to destroy a variety of contaminants, including TCE, perchloroethylene (PCE), trichloroethane (TCA), trihalomethanes, chlorobenzene, polychlorinated biphenyls (PCBs), and chlordane. The process could be carried out by trickling wastewater through a bed of iron and sand to give suitable retention and flow properties, or by fluidizing a bed of iron powder with the aqueous influent. Sweeny suggested that the reduction proceeds primarily by the removal of the halogen atom and its replacement by hydrogen (Equation 1-1), although other mechanisms probably play a role. Another important reaction suggested was the replacement of a halogen by a hydroxyl group (Equation 1-2). The iron metal was also believed to be consumed by water (Equation 1-3), although this reaction proceeds much more slowly.

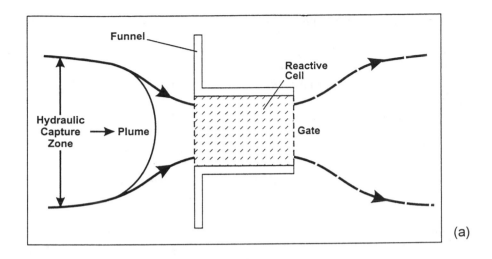

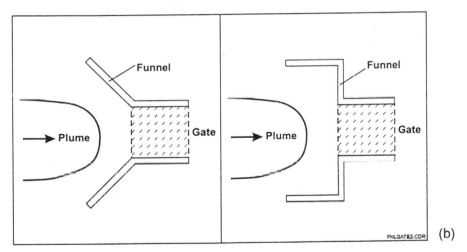

**FIGURE 1-2. (a) Funnel-and-gate system with straight funnel.
(b) Other possible funnel-and-gate system configurations**

$$Fe + H_2O + RCl \rightarrow RH + Fe^{2+} + OH^- + Cl^- \qquad \text{Eqn. 1-1}$$

$$Fe + 2H_2O + 2RCl \rightarrow 2ROH + Fe^{2+} \ 2Cl^- + H_2 \qquad \text{Eqn. 1-2}$$

$$Fe + 2H_2O \rightarrow Fe^{2+} + 2OH^- + H_2 \qquad \text{Eqn. 1-3}$$

Other researchers, such as Senzaki and Kumagai (1988a, 1988b) and Senzaki (1988) also suggested the use of iron powder for removal of TCE and TCA from wastewater. However, the significant potential of these discoveries and their in situ application was not realized until researchers at the University of Waterloo (Reynolds et al., 1990; Gillham and O'Hannesin, 1992) conducted focused efforts in this area. The University of Waterloo currently holds the patent for the use of zero-valent metals for in situ groundwater treatment (Gillham, 1993), and has granted commercialization rights to Envirometal Technologies, Inc. (ETI), a company partly owned by the University.

The exact mechanism of degradation of chlorinated compounds by iron or other metals is not fully understood. In all probability, a *variety of pathways* are involved, although recent research seems to indicate that certain pathways predominate. If some oxygen is present in the groundwater as it enters the reactive iron cell, the iron is oxidized and hydroxyl ions are generated (Equation 1-4). This reaction proceeds quickly, as evidenced by the fact that both the dissolved oxygen and the redox potential drop quickly as the groundwater enters the iron cell. The importance of this reaction is that oxygen can quickly corrode the first few inches of iron in the reactive cell. Under highly oxygenated conditions, the iron may precipitate out as ferric oxyhydroxide (FeOOH) or ferric hydroxide [$Fe(OH)_3$], in which case the permeability could potentially become considerably lower in the first few inches of the reactive cell at the influent end. Therefore, the aerobicity of the groundwater can be potentially detrimental to the technology. However, contaminated groundwater at many sites is not highly oxygenated. Also, engineering controls (see Chapter 3, Section 3.1.1.1) may be used to reduce or eliminate oxygenation in the groundwater before it enters the reactive cell.

$$2\ Fe^0 + O_2 + 2\ H_2O \rightarrow 2\ Fe^{2+} + 4OH^-$$ Eqn. 1-4

Once oxygen has been depleted, the reducing conditions created lead to a host of other reactions. Chlorinated organic compounds, such as TCE, are in an oxidized state because of the presence of chlorine. Iron, a strong reducing agent, reacts with the chlorinated organic compounds through electron transfers, in which ethene and chloride are the primary products (Equation 1-5).

$$3Fe^0 \rightarrow 3Fe^{2+} + 6e^-$$

Eqn. 1-5

$$\frac{C_2HCl_3 + 3H^+ + 6e^- \rightarrow C_2H_4 + 3Cl^-}{3Fe^0 + C_2HCl_3 + 3H^+ \rightarrow 3Fe^{2+} + C_2H_4 + 3Cl^-}$$

In one study, Orth and Gillham (1996) found that ethene and ethane (in the ratio 2:1) constitute over 80 percent of the original equivalent TCE mass. Partially dechlorinated byproducts of the degradation reaction, such as *cis*-1,2-dichloroethene (*c*DCE), *trans*-1,2-dichloroethene (*t*DCE), 1,1-dichlorothene, and vinyl chloride (VC), were found to constitute only 3 percent of the original TCE mass. Additional byproducts included hydrocarbons (C1 to C4) such as methane, propene, propane, 1-butene, and butane. Virtually all the chlorine in the original TCE mass was accounted for as inorganic chloride in the effluent, or as chlorine remaining on the partially dechlorinated byproducts. Similar results were obtained by Sivavec and Horney (1995), who quantified both liquid and gas phases of the reaction to obtain a carbon balance greater than 90 percent. A useful representation of the various proposed pathways is provided by Gillham (1996) and is reproduced here in Figure 1-3.

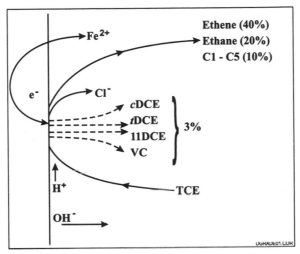

FIGURE 1-3. Schematic of proposed degradation process for TCE

A number of interesting issues are raised by this explanation of the reaction mechanism. For Equation 1-5 to take place in one step, without the generation of larger amounts of partially dechlorinated products (e.g., DCE or VC), six electrons have to be transferred almost instantaneously. Given the low probability of an instantaneous transfer of this magnitude, Orth and Gillham (1996) suggest that the TCE molecule must remain attached to the metal surface long enough for the six-electron transfer to occur. The TCE molecule remains attached to the metal surface either through the inherent hydrophobicity of TCE or, as Sivavec and Horney (1995) suggest, by the formation of a strong chloroethene-iron π-bond. This bonding prevents desorption until dechlorination is complete, although a few random chloroethene molecules may desorb early, leading to the presence of small amounts of DCE and VC. All this suggests that degradation of chlorinated organics by metals is a surface phenomenon and that the rate is governed by the specific surface area of the reactive medium.

Equation 1-5 shows the transformation of TCE to ethene, which is the primary product in many studies. It is unclear whether the other significant product, ethane, represents a different degradation pathway for TCE or whether it results from the iron-mediated catalytic transformation of ethene. Also unclear is whether the C1 to

C5 hydrocarbons represent an alternative pathway for TCE degradation or some other reaction. One study (Hardy and Gillham, 1996) suggests that aqueous CO_2 is reduced on the iron surface to form these hydrocarbon chains. Another study (Deng et al., 1996) suggests that the source of these hydrocarbons is the acid dissolution of gray cast irons containing both carbide and graphite carbon.

There is some indication that PCE and TCE in contact with iron may, at least partly, degrade through a different pathway from the one discussed above. Roberts et al. (1996) have seen indications of a reductive β-elimination pathway, in which PCE and TCE are reduced to dichloroacetylene and chloroacetylene, respectively. Both of these byproducts are potentially toxic, but are likely to be short-lived. Hydrogenolysis leads to their transformation to lesser chlorinated acetylenes, which are further reduced to substituted ethylenes. Hydrolysis of the chloroacetylenes to acetates is another possible pathway. These alternative pathways also could explain why only a small percentage (usually 3 to 5 percent) of the original PCE or TCE is realized as DCE and VC byproducts. In summary, there may be multiple pathways by which chlorinated ethylenes, such as PCE and TCE, are transformed in the presence of iron into dehalogenated products such as ethene.

Iron also reacts with *water* under reducing (anaerobic) conditions, although this reaction is believed to be much slower. The slow reaction with water (Equation 1-6) is advantageous to the technology because very little reactive medium (iron) is used up in this side reaction. Hydrogen gas and OH^- are formed as water is reduced, as shown in Equation 1-6.

$$Fe^0 \rightarrow Fe^{2+} + 2e^-$$

Eqn. 1 - 6

$$\frac{2H_2O + 2e^- \rightarrow H_2 + 2OH^-}{Fe^0 + 2H_2O \rightarrow Fe^{2+} + H_2 + 2OH^-}$$

Hydrogen generation could be a concern if hydrogen accumulates in the aquifer as it appears to do in some column tests. However, although initial field users were prepared to install hydrogen-gas collection systems, this has not been necessary at any site so far. Hydrogen generated by the slow reaction is believed to degrade through biological transformations.

Because several of the above reactions produce OH^-, the pH of the water in the reactive iron cell typically increases, often reaching values above 9.0. One effect of increased pH initially was thought to be a slowing down of the TCE degradation rate (O'Hannesin, 1993). The anticipation was that changes in pH might cause changes in the degradation rate through direct involvement of H^+ in Equation 1-5. However, subsequent research has cast doubt on whether or not pH affects degradation rate (Agrawal and Tratnyek, 1996).

An indirect effect of increasing pH is the potential for formation of precipitates, which could coat the surface of the iron and potentially reduce the reactivity of the iron and the hydraulic conductivity of the reactive cell. The dissolved carbonic acid and bicarbonate (alkalinity) present in many natural groundwaters act as buffers limiting pH increase and precipitate formation (Equations 1-7 and 1-8).

$$H_2CO_3^0 + 2\ OH^- \rightarrow CO_3^{2-} + 2\ H_2O \qquad \text{Eqn. 1-7}$$

$$HCO_3^- + OH^- \rightarrow CO_3^{2-} + H_2O \qquad \text{Eqn. 1-8}$$

Soluble carbonate ions are formed as the OH^- ions are consumed. If carbonate ions continue to build up, however, precipitation of carbonate solid species may occur. Depending on the composition of the groundwater, the precipitates formed could be calcite ($CaCO_3$), siderite ($FeCO_3$), or magnesium hydrocarbonates (Reardon, 1995). If groundwater carbonate is exhausted through the precipitation of carbonate minerals, the water may become saturated with respect to $Fe(OH)_2$ as the iron continues to oxidize. $Fe(OH)_2$ is relatively insoluble and may precipitate out if it builds up.

Most groundwaters contain many different aqueous species that may play some role in affecting the performance of a permeable barrier. In general, the course of chemical reactions taking place in multicomponent systems cannot be predicted by considering each species individually, because most of these reactions are interdependent. To some extent, equilibrium behavior in complex systems can be predicted using geochemical modeling codes, which are described in Chapter 5 (Section 5.2). However, many groundwater reactions in permeable barrier systems may not reach equilibrium during the passage of groundwater through the reactive cell. Another difficulty is that even if the type and mass of reaction products could be predicted, it is uncertain as to how many of these products are actually retained in the reactive cell and affect performance. For example, very fine precipitates that may be formed could be carried out of the reactive cell by colloidal transport with the groundwater flow. However, the reaction chemistry discussed above and the geochemical modeling codes described in Section 5.2 do provide a basis for selecting appropriate reactive media and planning performance monitoring schemes after permeable barrier installation.

1.4 POTENTIAL BIOLOGICALLY MEDIATED REACTIONS IN THE REACTIVE CELL

Gillham and O'Hannesin (1994) conducted column tests on TCE-contaminated water, both with and without an added biocide (azide). Similar degradation rates were observed in both cases, demonstrating that the degradation of TCE was abiotic and could proceed without microbial intervention. Also, microbial analysis (phospholipid fatty acid measurements) of the groundwater conducted during a pilot study at an industrial facility in Sunnyvale, California indicated no signs that the reactive media encouraged development of a microbial population over that in the surrounding aquifer (ETI, 1995).

However, the potential for microbially mediated processes in the reactive cell may be present under certain conditions. No significantly enhanced microbial activity has been noticed in the reactive cell at field installations to date (ETI,

1997). Appendix A, Section A-3 describes some possible biologically mediated reactions that have been proposed, given the geochemistry of groundwater and the presence of iron.

1.5 CURRENT STATUS OF PERMEABLE BARRIER FIELD APPLICATIONS

Table 1-2 describes the current status of pilot- and full-scale field applications of this technology at various sites. Granular iron was used as the reactive medium in all these field applications. The reactive medium was emplaced either in an excavated trench or a caisson driven into the ground. Installations were configured either as continuous reactive barriers or as funnel-and-gate systems. Funnel walls were either sealable-joint sheet piling or slurry walls. Trench-type reactive cells have been placed at depths down to 30 feet maximum. Caisson-type reactive cells have been placed at depths down to 40 feet maximum. At the present time, depth is the greatest limitation for field application of this technology.

1.6 EFFORTS OF GOVERNMENT AGENCIES TO PROVIDE REGULATORY GUIDANCE FOR THE USE OF PERMEABLE BARRIERS

In an effort to promote more regular consideration of newer, less costly, and more effective technologies to address the problems associated with hazardous waste sites, the U.S. EPA has published six In Situ Remediation Technology Status Reports, one of which deals with permeable barriers (U.S. EPA, 1995). This document briefly describes demonstrations, field applications, and research on permeable barriers. There is a growing interest in this technology at the state level as well.

The Interstate Technology and Regulatory Cooperation (ITRC) work group is striving to build a consensus on regulatory issues surrounding various remedial technologies. The ITRC has formed a Permeable Barriers Subgroup. This Subgroup first convened at a meeting in Philadelphia on September 25, 1996. The Permeable Barriers Subgroup includes members from environmental regulatory agencies in 29 states, as well as other interested parties such as environmental groups, the U.S. military, industry, and environmental consulting firms. The Subgroup will attempt to develop a consensus document for the states that will enhance the regulatory acceptance of this new technology and provide a consensus on requirements for compliance monitoring. Although a general regulatory consensus on permeable barriers will be announced by the Subgroup, individual states may decide to add on their own specific requirements.

According to the Subgroup, there currently are no specific regulatory requirements for installation or monitoring of permeable barriers (Turner, 1996). For the few systems that have been installed to date, regulatory requirements have been

TABLE 1-2. Current status of permeable barrier applications for chlorinated solvent sites

	Intersil, Sunnyvale, California	Industrial Facility, Upstate New York	Industrial Facility, Mountain View, California	Electronics Facility, Belfast, Northern Ireland	Industrial Facility, Coffeyville, Kansas
Pilot/Full Scale	Full	Pilot	Full	Full	Full
Date Installed	12/94-1/95	5/95	9/95	12/95	1/96
Type of Barrier	Iron gate with slurry wall funnels	Iron gate with sheet pile (funnels ~5 ft either side)	Continuous iron cell	In situ caisson reactor with slurry walls (80-100 ft on either side)	Iron cell with slurry walls (500 ft on either side)
Depth of Barrier	20 ft	15 ft	25 ft	40 ft	28 - 30 ft
Type of Media	Iron	Iron	Iron	Iron	Iron
Total (and Unit) Cost of Media	$170,000 ($650/ton)	$30,000 ($650/ton)	$60,000 ($650/ton)	$20,000 ($450/ton)	$50,000 ($650/ton)
Contaminants and Concentrations	TCE, cDCE, VC, Freon™-113	TCE 30-380 ppb VC 4.9-7.1 ppb cDCE 98-550 ppb 1,1,1-TCA 3.2-13 ppb	cDCE 5-10 ppm VC 5-50 ppb TCE up to/1 ppm	TCE 300,000 ppb 1,1,2-TCA 200 ppb cDCE 2000 ppb Trace others ppb	TCE 400 ppb
Emplacement Technique Used for Funnel-and-Gate Portions	Gate-trench box Funnel-slurry walls	Gate-trench box Funnel-sealable sheet pile	Emplaced by backhoe during backfilling of open pit	Cylindrical reactive vessel filled with iron, slurry walls	Soil-bentonite slurry walls
Installation Cost (all U.S. dollars)	$600,000	$220,000	Completed as part of overall site remediation (not a big incremental cost over backfilling)	$315,000	$400,000 including iron
Compliance Monitoring Well Locations	In last 4 inches of iron cell	In downgradient pea gravel	In downgradient section of iron	Various locations in cylinder	Various locations in iron cell
Comments	MODFLOW model	FLOWPATH model	No computer modeling	FLOWPATH model	FLOWPATH model

TABLE 1-2. Current status of permeable barrier applications for chlorinated solvent sites (page 2 of 3)

	Lowry AFB, Denver, Colorado	Moffett Federal Airfield, California	USCG Facility, Elizabeth City, North Carolina	Canadian Forces Base Borden, Canada	Denver Federal Center, Colorado
Pilot/Full Scale	Pilot	Pilot	Full	Pilot	Full
Date Installed	11-12/95	4/96	6/96	6/91	10/96
Type of Barrier	Funnel-and-gate, sheet pile walls	Funnel-and-gate, sheet pile walls	Continuous iron trench (150 ft)	Continuous reactive wall	Funnel-and-gate 1,000 ft long, four 40-ft gates
Depth of Barrier	18 ft	25 ft	24 ft	32 ft	20 ft
Type of Media	Iron	Iron	Iron	Iron and sand mixture	Iron
Total (and Unit) Cost of Media	$32,500 ($650/ton)	NA	±$171,000 ($380/ton)	NA	($375-400 ton)
Contaminants and Concentrations	TCE 1,000 ppb cDCE 250 ppb VC 25 ppb	TCE >20 mg/L PCE 0.5 mg/L	TCE up to 16 mg/L Cr 6-10 ppm	TCE 250,000 µg/L PCE 43,000 µg/L TCM NA	
Emplacement Technique Used for Funnel-and-Gate Portions	Sealable-joint sheet pile funnel walls; trenched gate	Sealable-joint sheet pile funnel walls; trenched gate	Continuous trencher	Clamshell excavated trench	Sheet pile for funnel
Installation Cost (all U.S. dollars)	$137,500 including iron	$380,000 including iron	$350,000 including iron	NA	$900,000-$1,000,000 (final not yet known)
Compliance Monitoring Well Locations	Various locations in iron	Various locations in iron	Upgradient and downgradient	2 wells upgradient, 3 downgradient, 6 in iron	Various locations in iron and downgradient pea gravel
Comments	MODFLOW model	MODFLOW model	FLOWPATH	NA	MODFLOW model

TABLE 1-2. Current status of permeable barrier applications for chlorinated solvent sites (page 3 of 3)

	Industrial Facility, New Jersey	Somersworth Sanitary Landfill, New Hampshire	Alameda Naval Air Station, California
Pilot/Full Scale	Pilot	Pilot	Pilot
Date Installed	11/94	Current (10-11/96)	11-12/96
Type of Barrier	Aboveground iron reactor	Slurry walls (10 ft), 8-ft-diameter caisson with iron	Sheet pile with sequential iron, O_2 sparge gate
Depth of Barrier	NA	40-45 ft	15 ft
Type of Media	Iron	Iron	Iron + O_2 sparge gate
Total (and Unit) Cost of Media	NA	~ $100,000 total	Iron $375/ton
Contaminants and Concentrations	TCE 3,000 µg/L PCE 50,000 µg/L	TCE 310 ppb PCE 3.7 ppb cDCE 565 ppb VC 387 ppb	cDCE VC BTEX 1-5 mg/L
Emplacement Technique Used for Funnel-and-Gate Portions	NA	Caissons slurry walls	Trench box
Installation Cost (all U.S. dollars)	NA	$175,000	To be determined
Compliance Monitoring Well Locations	Side ports along reactor	Various locations in iron	Various locations in iron
Comments	NA	Visual MODFLOW, FRAC3DVS	Rice/U. of Waterloo Project Visual MODFLOW

NA = not available.
BTEX = benzene, toluene, ethylbenzene, and xylenes.

determined on a case-by-case basis. The California Department of Toxic Substances Control likewise makes its determinations on a case-by-case basis (Hadley, 1996). Regulatory agencies currently suggest that for a prospective site there should be (1) compelling reasons why a permeable barrier is the best choice for that site and (2) data to show why it is expected to work as planned. Different states, or even different agencies within a state, are likely to have different requirements.

Intersil, the site with the first full-scale permeable barrier application, was in many ways an ideal situation from a technical feasibility and regulatory viewpoint. It was an underutilized property, was run by a cooperative potentially responsible party (PRP), and posed no excessive human health threat. Further, it had shallow groundwater, poor (brackish) water quality, a competent aquitard, and a wall that was not too deep. A pilot study on site showed that the barrier would work and that the total cost was estimated as half that of a pump-and-treat system over 30 years (Kilfe, 1996). The cost analysis for this site assumed that the iron would not require replacement and included the benefit of being able to lease the property, an option that was enabled by the passive long-term nature of the technology. Although the plume had moved off the property at Intersil, regulators allowed placement of the permeable barrier within property lines based on indications that natural attenuation of the chlorinated contaminants was occurring downgradient.

Other sites may be more difficult from a regulatory viewpoint. At one potential site where a full-scale permeable barrier is being considered, the approval process is made difficult by the fact that there is already a Record of Decision (ROD) with 30 signatories (PRPs) in place for installing a pump-and-treat system to clean up a regional plume. There may also be problems with site access if the plume has already moved beyond the property boundary and the permeable barrier is required to be installed at the leading edge of the plume.

One important trend is that regulators are increasingly open to discussion of cleanup costs. There is a growing willingness in the regulatory community to consider cost an important factor in selecting alternatives for cleanup. If a large benefit-to-cost ratio can be shown for the permeable barrier versus a pump-and-treat system, it would be a considerable factor in favor of its selection.

For many chlorinated compounds, the degradation potential of a permeable iron barrier has been demonstrated with a fair degree of confidence in a variety of published research. Site-specific degradation feasibility can be shown through the treatability tests described in Chapter 4.0. From a regulatory perspective, the questions that remain relate to ensuring hydraulic capture of the plume, maintaining long-term performance (in terms of hydraulics and reactivity), and verifying downgradient water quality. These issues can be addressed in part by developing a site-specific monitoring program for the permeable barrier system. It is recommended that site managers confer with regulators as early as possible in the design stage to promote better understanding of the needs of all concerned parties.

CHAPTER 2

SITE CHARACTERIZATION DATA

In comparison with a pump-and-treat system, a permeable barrier is a relatively permanent structure. Whereas for a pump-and-treat system, the locations of pumping wells, pumping rates, and aboveground treatment methods can be changed or modified as understanding of the site grows, a permeable barrier is difficult to relocate and change. Therefore, it is important to understand the site as well as possible before installing the barrier. The following aspects of the site are important to know:

- Groundwater flow system characteristics
- Organic composition of the groundwater
- Inorganic composition of the groundwater.

Because seasonal variations in such factors as flow and rainfall events could affect these site parameters, quarterly data collected over the period of 1 year are desirable. Some parameters may be sampled more frequently or continuously for added certainty. However, at many sites, designers have to work with what is available. If data appear to be inadequate, it may be possible to do some additional characterization while collecting groundwater samples for laboratory treatability testing. Section A.1 in Appendix A contains an example of a site profile sheet that could be used to determine the suitability of various sites for permeable barrier treatment and to establish treatability testing and modeling parameters.

2.1 GROUNDWATER FLOW SYSTEM CHARACTERISTICS

The requirements for groundwater flow system characterization include data on geologic and hydrologic parameters. In most cases, the needed information is available from previous studies conducted at the sites, such as the Remedial Investigation/Feasibility Study (RI/FS), ROD reports, and groundwater modeling reports. Sometimes, additional site-specific characterization may be needed to support the feasibility study, site selection, and design of the permeable barriers. The example preliminary site profile questionnaire (Section A.1) asks for brief data on site stratigraphy, soil types, depth of water, groundwater flow direction, groundwater velocity, hydraulic conductivity (K), porosity, depth to confining layer, and dimensions and depth of the dissolved plume. However, once the decision to implement the permeable barrier technology has been taken based on a preliminary assessment of site data, more detailed information on these and other parameters may be needed during the design phase of the process. Various aspects of the detailed site hydrogeologic characterization and their implications are discussed in Section A.2 and summarized below.

2.1.1 Site Background Information

General background information on the physical features of the site and the groundwater regime is needed for initial assessments of the feasibility of a permeable barrier at the site candidate locations (as well as for establishing whether additional site characterization is necessary). Background information generally is available in existing documents, such as previous site characterization reports, aerial photographs, site maps, well logs, underground utility maps, and quarterly monitoring reports. Surface and subsurface features that may impede the installation of an in situ barrier usually are identified at this stage.

2.1.2 Hydrostratigraphic Framework

The hydrostratigraphic framework involves the site-specific geologic and hydrologic data required to construct a conceptual understanding of groundwater flow at the site. This may include information collected from previous reports, as well as new information collected specifically to evaluate and design a permeable barrier. The most significant data to be collected include variations in the depth, thickness, and water levels of different hydrostratigraphic units (HSUs). This is achieved by drilling and sampling several locations by conventional drilling or other techniques, such as cone penetrometer testing (CPT) or the use of a GeoProbe™. The number and locations of boreholes and samples required for the site heterogeneity assessment should be based on the scientific judgment of the on-site hydrogeologist and the availability of pre-existing data.

2.1.3 Hydrologic Parameter Estimation

Hydrologic or groundwater flow parameters are very important in permeable barrier design. The following parameters greatly influence the configuration and dimensions of the barrier installed and are discussed in more detail in Section A.2:

- Hydraulic conductivity (K)
- Porosity
- Hydraulic gradient
- Groundwater flow direction and velocity.

2.2 ORGANIC COMPOSITION OF THE GROUNDWATER

Information on the extent and type of chlorinated organics contamination is critical for the feasibility study, batch and column tests, design development, permeable barrier placement, and performance monitoring. The presence and concentration of nonchlorinated organic compounds also may be of interest in selecting an appropriate reactive medium or treatment scheme.

2.2.1 Organic Contaminant Spatial Distribution

This section deals with the spatial distribution of chlorinated organics contamination. In many cases, several different contaminant plumes are present at a single site. In general, the three-dimensional distribution of each contaminant plume at the site needs to be delineated so that the permeable barrier can be appropriately sized to capture it. This includes the identification of the contaminated aquifer(s), the depth and width of the plume(s), the average and maximum concentration, and the rate of plume movement. In addition, it is important to characterize the significant processes that affect the spread of contamination in the subsurface at the site. These may include the effects of adsorption/retardation, chemical reactions, dispersion, and vertical plume movement due to fluid density effects.

In many cases, some of the required data already is available from the RI/FS, ROD, Resource Conservation and Recovery Act (RCRA) Facility Investigation (RFI), or routine monitoring reports from the site. Therefore, there may be no need to acquire new data for plume characterization at these sites. Instead, a careful review of existing reports should be conducted and new data should be collected only if significant data gaps are found or if the preexisting data are out of date or inadequate. If needed, groundwater samples can be collected to fill data gaps for specific analytes or to improve sampling density in areas of particular interest. Unless available from the recent site reports, the compiled data should be plotted on isopleth maps and on the cross-sectional profiles to evaluate the lateral and vertical extent of the plumes.

The width of the contaminant plume can be determined from the isopleth maps of concentration. If sufficient data are available, the maps also may reveal the potential source zones for the contaminants and the existence of preferential pathways for contaminant migration along which the contaminants have advanced. The plume maps also can be used to identify a potential location and design for the permeable barrier installation. In most cases, the barrier is installed near the downgradient end of the plume. However, several factors may lead to the installation of barriers within the plumes. For example, barriers may have to be installed at the property boundary, even if a portion of the plume already has moved past the property boundary, if site access to the edge of the plume is difficult. Sometimes the barrier may be located in the proximity of the highest concentration parts of the plume to expedite the remediation of the most contaminated areas. Such a location may be required for slow moving plumes. Other measures may be required in such cases to address the remaining portion of the plume.

The design of the permeable barrier is controlled partly by contaminant distribution. In most cases, the barrier should be installed to capture the entire plume width and depth. This may be done either by installing a reactive barrier across the entire width or by installing a small reactive cell flanked by funnel walls. These design considerations for the permeable barriers are further discussed in the modeling information in Chapter 5 (Section 5.1).

2.2.2 Groundwater Sampling for Volatile Organic Compounds (VOCs)

Groundwater sampling provides essential information on water movement, contaminant levels, and inorganic chemistry and geochemistry needed to understand and model the performance of the reactive cell.

Proper quality assurance (QA) procedures, as described in EPA SW-846 (EPA, 1994), should be followed during sampling to ensure valid data. Zero headspace should be ensured prior to sealing the sample containers. Sample containers should be labeled, logged, and stored at approximately 4°C while they are being transferred under chain-of-custody protocol to an analytical laboratory for analysis. Analysis must be completed prior to expiration of recommended holding times. Field duplicates, field blanks, and trip blanks are commonly used quality control (QC) samples that aid data quality evaluation.

2.2.3 Analytical Methods for VOCs

This section briefly describes the methods used for analysis of groundwater to meet the essential requirements of a site characterization study. VOCs in groundwater samples can be analyzed by EPA Method 8240 (*Volatile Organic Compounds by Gas Chromatography/Mass Spectrometry* [GC/MS]), EPA SW-846, Update II, September 1994) or EPA Method 8260 (similar to Method 8240, but uses capillary column) in conjunction with EPA Method 624 (U.S. EPA, 1991). Method 624 is a sample preparation and extraction procedure for analysis of VOCs by a purge-and-trap apparatus. This technique can be used for most VOCs that have boiling points below 200°C and are insoluble or slightly soluble in water. Volatile, water-soluble compounds can be included in this analytical technique; however, quantitation limits by gas chromatography (GC) are generally higher because of poor purging efficiency.

QA involves the use of blanks, duplicates, and matrix spikes to ensure laboratory data quality. The accuracy and precision of Method 8240 or Method 8260 are related to the concentration of the analyte in the investigative sample and are essentially independent of the sample matrix. Linear equations pertaining to accuracy and precision for a few compounds are discussed in the standard method descriptions. The estimated quantitation limit (EQL) for individual compounds is approximately 5 μg/L in groundwater samples. EQLs are proportionally higher for sample extracts and samples that require dilution or reduced sample size to avoid saturation of the detector.

2.3 INORGANIC COMPOSITION OF THE GROUNDWATER

Monitoring of inorganic field parameters such as pH, redox potential (Eh), and dissolved oxygen (DO) in the groundwater is very important because they can be used to determine whether conditions at the site are conducive to formation of inorganic precipitates. These three groundwater field parameters should be monitored over at least 1 year to evaluate seasonal fluctuations. Similarly, chemical species that may react in the conditions created by the reactive medium include Ca, Fe, Mg, Mn, Al, Ba, Cl, F, SO_4^{2-}, and HCO_3^- (alkalinity); significant redox-sensitive elements include Fe, Mn, C, S, and N. For example, iron in solution may be in the ferrous [Fe(II)] state or ferric [Fe(III)] state, and organic carbon as humic or fulvic substances may be reduced to methane in the reactive cell. Sulfate [S(VI)] may be reduced to bisulfide [S(−II)] and nitrate may be reduced to nitrogen gas [N(0)] or ammonia [N(−3)] if conditions are conducive. Bromide is measured mainly because it potentially can be used as a tracer during performance monitoring of the barrier.

2.3.1 Sampling and Analysis of Field Parameters

The primary purpose of taking field parameter measurements is to monitor aquifer conditions that can affect the performance of the reactive wall. Therefore, the water level, temperature (T), pH, Eh, and DO should be measured at designated monitoring wells. To obtain accurate readings, T, pH, and Eh should be measured using the most appropriate method available to provide representative values. Typical devices include a downhole probe with multiple sensors and a flowthrough cell shielded by an inert gas. Other parameters, such as specific conductance, turbidity, and salinity of a groundwater sample can be measured ex situ, if required, using appropriate field instruments. Table 2-1 lists the field parameters and corresponding analysis methods.

2.3.2 Sampling and Analysis for Inorganic Chemical Parameters

Inorganic analytes should be measured because they provide valuable information about the demands on the reactive medium itself. Samples should be collected from each monitoring point for laboratory analysis as indicated in Table 2-1. Samples should be filtered (for cations only) and preserved immediately after collection. The typical filter pore size for cation analysis is 0.45 μm; however, filters of different pore size may be used from time to time for comparison. In addition, several samples should be collected and preserved without filtering to determine the content in the suspended matter. Total dissolved solids (TDS) and total suspended solids (TSS) should be determined from filtered and unfiltered samples, respectively. QA procedures include the use of blanks, duplicates, and matrix spikes to ensure data quality.

TABLE 2-1. Requirements for field parameters and inorganic analytes (based on EPA SW-846)

Analyte or Parameter	Analysis Method	Sample Volume	Storage Container	Preservation	Sample Holding Time
Field Parameters					
Water level	In-hole probe	None	None	None	Minimal
pH	In-hole probe	None	None	None	Minimal
Groundwater temperature	In-hole probe	None	None	None	Minimal
Redox potential	In-hole probe	None	None	None	Minimal
Dissolved oxygen	In-hole probe	None	None	None	Minimal
Specific conductance	Field instrument	Minimal	None	None	Minimal
Turbidity	Field instrument	Minimal	None	None	Minimal
Salinity	Field instrument	Minimal	None	None	Minimal
Inorganic Analytes					
Metals (K, Na, Ca, Mg, Fe, Al, Mn, Ba)	200.7	100 mL (all)	Polyethylene	Filter[a], 4°C, pH<2 (HNO_3)	180 days
Anions (NO_3, SO_4, Cl, Br, F)	300.0	100 mL (all)	Polyethylene (all)	4°C (all)	28 days (48 hr for NO_3)
Alkalinity	310.1	100 mL	Polyethylene	None	14 days
Other					
TDS, TSS	160.2, 160.1	100 mL	Polyethylene	4°C	7 days
TOC, DOC	415.1	40 mL	Glass	4°C, pH <2 (H_2SO_4)	7 days

(a) 0.45 μm pore size.
TDS = total dissolved solids.
TSS = total suspended solids.
TOC = total organic carbon.
DOC = dissolved organic carbon.

CHAPTER 3

REACTIVE MEDIA SELECTION

Once site characterization information has been obtained, a suitable reactive medium has to be selected for use in the reactive cell. The choice among reactive metal media for the reactive cell is governed by the following considerations:

- **Reactivity.** A medium that affords lower half-lives (faster degradation rates) is preferred.

- **Stability.** Length of time that a reactive medium or that mixed media will maintain reactivity is an important concern. No full- or pilot-scale barrier has been operating for a sufficient length of time to make a direct determination of stability. However, an understanding of the reaction mechanism can provide some indication of the future behavior of the medium.

- **Availability and Cost.** A cheaper medium is preferred over a more expensive medium, especially if reported differences in performance are slight.

- **Hydraulic Performance.** The particle size of the reactive medium should be sufficient to ensure required hydraulic capture by the barrier.

- **Environmental Compatibility.** The reactive medium should not introduce harmful byproducts into the downgradient environment.

There may be a trade-off between these factors, and final selection may have to be based on the importance of each factor for a given site.

3.1 TYPES OF REACTIVE MEDIA AVAILABLE

Several different types of reactive metal media are available for use in permeable barriers and are discussed below.

3.1.1 Granular Zero-Valent Metal

Granular zero-valent metal, particularly iron, is the most common medium used so far in bench-, pilot-, and full-scale installations.

3.1.1.1 Granular Iron. Both reagent- and commercial-grade iron have provided significant rates of chlorinated solvent reduction in water. Sivavec and

Horney (1995) studied the degradation rates for chlorinated compounds with commercial iron from 25 different sources. These, as well as other researchers (Agrawal and Tratnyek, 1996; Matheson and Tratnyek, 1994), have found that the primary determinant of degradation rate in different irons is the available reactive surface area. The parameter generally used to discriminate between different irons is the specific surface area, or the surface area per unit mass (m^2/g) of iron.

Sivavec and Horney (1995) found that pseudo first-order degradation kinetics (with respect to chlorinated ethene concentrations) were applicable when the ratio of iron surface area to volume of aqueous phase ranged from 0.1 to 1,325 m^2/L. The surface area of the metal was measured by the Brunauer-Emmett-Teller Adsorption Isotherm Equation (BET) Kr or N_2 adsorption. Specific surface areas of untreated iron from the 25 different sources varied over four orders of magnitude. Acid pretreatment was found to increase the degradation rate of iron (Agrawal and Tratnyek, 1996; Sivavec and Horney, 1995), probably due to removal of any passivating oxide layer on the iron or due to an increase in the surface area by etching or pitting corrosion. Therefore, commercial irons with higher surface area are preferred. However, the higher surface area requirement for reactivity should be balanced with the hydrogeologic necessity to select a particle size that affords a reactive cell hydraulic conductivity that is at least five times (or more) higher than that of the surrounding aquifer (see Chapter 5, Section 5.1.3). Generally, sand-sized particles of iron are selected for use in reactive cells. The hydraulic conductivity of the reactive cell can also be improved by mixing sand (or coarser concrete sand) with finer iron particles. Adding sections of pea gravel along the upgradient and downgradient edges of the reactive cell also improves the distribution of flow through the reactive cell, and this feature has been used in several field installations to date.

Another variation of granular iron media under serious consideration by technology developers is the use of a pretreatment cell containing a coarse medium (sand or pea gravel) mixed with a small percentage (10 to 15 percent) of iron. This pretreatment cell would remove primarily dissolved oxygen from aerobic groundwater before it enters the reactive cell containing 100 percent iron. With its higher porosity, this coarse pretreatment zone is better able to handle the precipitates that are formed, compared to the reactive cell itself.

Based on experience at existing permeable barrier applications, the general requirements for the iron used in reactive cells are as follows (ETI, 1996):

- Iron ideally should be over 95 percent by weight of Fe^0, with minor amounts of carbon, a minimal oxide coating, and no hazardous levels of trace metal impurities. Many suppliers perform environmental quality testing on their materials to determine the concentration of impurities.

- A desirable grain-size range for many applications is between -8 to $+50$ mesh. Bulk density measurements may be provided by the supplier.

- Because the iron may be generated from cutting or grinding operations, it should be ensured that there are no residual cutting oils or grease on the iron.

- A Material Safety Data Sheet (MSDS) that identifies health and safety hazards of the material would be desirable.

3.1.1.2 Other Zero-Valent Metals. A number of other zero-valent metals have been investigated for their potential to reduce chlorinated hydrocarbons. Experiments were conducted to determine the relative rates of reduction of various hydrocarbons by stainless steel, Cu^0, brass, Al^0, mild steel, and galvanized metal (Zn^0) (Reynolds et al., 1990; Gillham and O'Hannesin, 1992). Mild steel and galvanized metal had the fastest reduction rates, followed by Al^0. Little reduction occurred with stainless steel, Cu^0, and brass. These results indicate that there would be no significant advantage to using any of these metals over Fe^0. Boronina et al. (1995) investigated the reactivity of Mg^0, Sn^0, and Zn^0 with CCl_4. Rapid oxidation of Mg^0 by water effectively prevented it from reducing CCl_4. Sn^0 and Zn^0 were capable of degrading CCl_4; however, the cost, the incomplete degradation of chlorinated reaction products, and the dissolution of these toxic metals must be considered before the use of these metals can be considered as a viable alternative to Fe^0. Schreier and Reinhard (1994) have investigated the ability of Fe^0 and Mn^0 powders to reduce several chlorinated hydrocarbons. Experiments conducted with manganese followed zero-order kinetics. The rates determined appeared to be fairly slow; the zero-order rate constants were determined to range from 0.07 to 0.13 molar units/day, depending on the aqueous-phase solution composition.

3.1.2 Granular Iron with an Amendment

Oxidation of Fe^0 to Fe^{2+} results in an increase in pH. Depending on a variety of physical and chemical factors (e.g., flowrate through the barrier and groundwater geochemistry), this increase in pH can result in the precipitation of a number of minerals, including $Fe(OH)_2$, $FeCO_3$, and $CaCO_3$. Various amendments can be added to the granular iron to moderate the pH. Pyrite has been used successfully in laboratory experiments for moderating the pH (Burris et al., 1995; Holser et al., 1995). The oxidation of the pyrite produces acid, which offsets the acid consumed during the oxidation of Fe^0:

$$Fe^0 + 2H^+ + \tfrac{1}{2} O_2 \Rightarrow Fe^{2+} + H_2O \qquad \text{Eqn. 3-1}$$

The net reaction for pyrite oxidation is as follows:

$$FeS_2 + {}^7\!/_2 O_2 + H_2O \Rightarrow Fe^{2+} + 2H^+ + 2SO_4^{2-} \qquad \text{Eqn. 3-2}$$

In addition to lowering the pH, the addition of pyrite and iron sulfide to Fe^0 has been shown in the laboratory to reduce the half-life of carbon tetrachloride

(Lipczynska-Kochany et al., 1994). At a ratio of FeS_2/Fe^0 of 0.03, the half-life of carbon tetrachloride was reduced by 6 percent over that of iron alone. At a FeS_2/Fe^0 ratio of 0.11 the half-life was reduced by 45 percent. Ferrous sulfide also reduced the half-life of carbon tetrachloride reduction by Fe^0. When added at a ratio (FeS/Fe) of 0.04, the half-life of carbon tetrachloride was reduced by 18 percent. In addition to the materials discussed above, other materials have been proposed for moderating the pH, including troilite, chalcopyrite, and sulfur. One side effect of adding pH-controlling amendments could be the potential for higher levels of dissolved iron in the water downgradient from the reactive cell.

3.1.3 Bimetallic Media

A number of bimetallic systems in which various metals are plated onto zero-valent iron have been shown to be capable of reducing chlorinated organic compounds at rates that are significantly more rapid than zero-valent iron itself (Sweeny and Fisher, 1973: Sweeny, 1983; Muftikian et al., 1995; Korte et al., 1995; Orth and McKenzie, 1995). Some bimetals (e.g., Fe-Cu) act as galvanic couples and enhance the degradation rate by increasing electron activity. Other bimetals, such as iron-palladium (Fe-Pd), enhance the degradation rate because Pd acts as a catalyst. Recent studies (Appleton, 1996) have shown that the Fe-Ni bimetallic system has the potential to considerably enhance reaction rates. There may be a cost trade-off between the construction of a smaller reactive cell (because of the faster reaction rate) and the higher cost (relative to granular iron) of the new highly reactive medium.

Of the bimetallic systems studied so far, the Fe-Pd bimetal appears to have the fastest reaction kinetics. Laboratory studies with palladized iron have demonstrated that the reduction of TCE can be increased by up to two orders of magnitude over that of iron alone (Muftikian et al., 1995; Korte et al., 1995; Orth and McKenzie, 1995). In addition, palladized iron allows for the reduction of some of the more recalcitrant compounds, such as dichloromethane. However, due to the high cost of palladium, this system is not likely to be cost effective in an in situ application.

One caution that should be exercised while examining bimetallic media in particular, and all media in general, is to ensure that the enhanced reactivity can be maintained over long periods of time. There is some preliminary indication from long-term column tests that the reactivity of bimetallic systems initially may be high, but may decline gradually after several pore volumes of groundwater have flowed through (Sivavec, 1997). Also, the metals used in bimetallic systems (e.g., Fe-Ni) should not introduce environmentally undesirable levels of dissolved metals into the downgradient aquifer.

3.1.4 Other Innovative Reactive Media

3.1.4.1 Cercona™ Iron Foam. One group of materials that appears promising for use in permeable barriers is ceramic foam and aggregate products of the type made by Cercona, Inc., Dayton, OH (Bostick et al., 1996). The iron foam material can potentially achieve high surface area (high reactivity) and high porosity at the same time, while still using zero-valent iron. In contrast, conventional reactive granular media have a trade-off between surface area and porosity. These iron foam materials are based on gelation of soluble silicates with soluble aluminates. These two solutions are combined with an aggregate or powdered material in a controlled and reproducible manner under specific conditions including the solution concentration, the temperature, and the ratio of materials to make the final product. The addition of the custom aggregate or powdered material to the silicate/aluminate slurry results in a final product with a composition that typically is 5 to 15 percent silicate and aluminate with the balance being the additive of choice. The additives are based on the desired properties of the product. For permeable reactive barrier applications, typical additives would include metallic iron, iron oxides, zeolites, clays, or specialty ceramic materials.

3.1.4.2 Colloidal Iron. Granular iron materials of sand size and larger have been the most common reactive media used in laboratory and field studies of permeable reactive barriers so far. One alternative form of iron that has been suggested is colloidal-size iron material (1 to 3 microns in diameter). This material is considerably more expensive than granular iron materials; however, it may have some advantages over granular materials. Colloidal-size iron allows the formulation of slurries that can be injected into the aquifer, making it possible to emplace a permeable reactive barrier anywhere a well can be installed, including in deep sites and fractured media. Some studies have explored the viability of this innovative approach for emplacement of an in situ permeable barrier composed of iron (Kaplan et al., 1996; Cantrell and Kaplan, 1996; Cantrell et al., 1997). In the proposed approach, colloidal-size iron particles would be injected as a suspension into the subsurface. As the suspension of particles moves through the aquifer material, the particles would be filtered out on the surfaces of the aquifer matrix. As a result of the high density of the iron particles (7.6 g/cm^3), it appears that the primary removal mechanism of iron colloids in aqueous solution passing through sand columns is gravitational settling. Because colloidal-size iron particles have higher surface areas, a lower total iron mass may be required in the treatment zone. Cantrell and Kaplan (1996) estimate that a 1.0-m-thick chemically reactive barrier with an iron concentration of 0.4 percent by volume would last for approximately 30 years under typical groundwater conditions. Although laboratory column experiments have been promising, this technology has not been tested in the field.

3.1.4.3 Ferrous Iron-Containing Compounds. In addition to zero-valent iron, several ferrous iron-containing compounds have been investigated for their potential as suitable reducing agents for chlorinated hydrocarbons. Lipczynska-Kochany et al. (1994) found that Na_2S, FeS, and FeS_2 all were capable of reducing carbon tetrachloride with half-lives that were nearly the same as Fe^0 (approximately 24 minutes). Kriegman-King and Reinhard (1991, 1994) also investigated the reduction of carbon tetrachloride by pyrite. The reaction rates that they observed appear to be similar; however, it is difficult to make an objective comparison because different experimental conditions were used in each study.

3.1.4.4 Reduction of Aquifer Materials by Dithionite.

Fe^{2+} appears to facilitate transformations of reducible organic substances in aqueous systems (Amonette et al., 1994; Sivavec et al., 1995). As a result, it has been proposed that reduction of ferric iron in aquifer minerals to produce surface-bound ferrous iron may be a promising in situ remediation technology (Amonette et al., 1994). In the proposed approach, a solution containing sodium dithionite $(Na_2S_2O_2)$ would be injected into the aquifer to reduce ferric iron in aquifer minerals to surface-bound ferrous iron. The surface-bound ferrous iron then acts as the reductant for the destruction of the chlorinated hydrocarbons. Laboratory experiments (batch) were conducted to determine the effectiveness of dithionite-reduced sediments for destruction of carbon tetrachloride (Amonette et al., 1994). The results indicate that the half-life of carbon tetrachloride in contact with dithionite-reduced sediments (at 30°C) is 48 hours. However, additional batch and column experiments have indicated that dithionite reduction of TCE is much less effective. The half-life determined for TCE was over 90 days. These results suggest that using dithionite-reduced aquifer materials may be a viable in situ remediation technique for easily reduced compounds such as carbon tetrachloride, CrO_4^{2-}, and UO_2^{2+}, but not for compounds that are more difficult to reduce such as TCE.

3.2 SCREENING AND SELECTION OF REACTIVE MEDIA

In general, suitable reactive media should exhibit the following properties:

- Sufficient reactivity to degrade the contaminants with an economically viable flowthrough thickness (residence time).

- Ability to retain this reactivity under site-specific geochemical conditions for an economically viable period of time (several years or decades).

- Sufficient particle size to create a porosity that allows the creation of a reactive cell that captures the targeted plume width.

- Ability to retain the porosity (hydraulic conductivity) at or above minimum specified levels over time, through the inhibition of precipitate formation under site geochemical conditions.

- Environmentally compatible reaction products (e.g., Fe^{2+}, Fe^{3+}, oxides, oxyhydroxides, and carbonates).

- Easy availability at a reasonable price.

Batch tests can be performed (see Section 4.1) to screen prospective media. Even granular iron from two or more sources may have to be tested to see which one works better. However, once the number of media have been narrowed to two or three candidates, column tests should be performed, as described in Chapter 4 (Section 4.2), to determine half-lives and select the final medium.

Geochemical models (see Chapter 5, Section 5.2) can also help identify candidate media by examining potential reaction products in the reactive cell. This kind of geochemical modeling is referred to as *forward or predictive modeling*, in which a set of reactions and their stoichiometries are assumed and the final outcome of water composition and mineral assemblage is calculated using a computer program. The initial state of the groundwater usually is taken as its composition prior to encountering the reactive medium. The final (equilibrium) composition of the water and the mass of mineral matter that is precipitated or dissolved depend somewhat on its initial chemical makeup. For example, groundwaters that are high in inorganic carbon, due to contact with carbonate minerals or to plant respiration along a root zone, may become oversaturated with respect to some minerals in the reactive cell due to an increase in dissolved iron and elevated pH. Such conditions can lead to precipitation of minerals and other solids within the reactive cell. Predictions based on this kind of modeling should be tested using batch or column experiments in a few cases, to verify that relevant system parameters are well understood and can be applied to a laboratory-scale design. If, based on predictive simulations or treatability testing, precipitation is thought likely to occur, a different medium or mixture of media may be tested experimentally or modeled using a forward geochemical code, and the results may be used to attempt to minimize the potential for precipitation.

Another application of geochemical modeling to media selection is *inverse modeling* (see Section 5.2), which calculates the outcome of probable reactions based on chemical data at initial and final points along a flowpath. The difference between inverse and forward modeling is that the former does not necessarily represent equilibrium. Rather, changes in groundwater composition are attributed to changes in solid precipitation or dissolution. Another important difference is that inverse modeling has the capability to predict the amount of mass change that must occur to satisfy the observed conditions, whereas in forward modeling only the tendency for such changes is determined. Inverse modeling as a tool for media selection is best utilized by predicting mass changes in a column experiment based on analysis of column influent and effluent or the groundwater at different locations within the column. Rates of reactions and subsequent mass changes then may be calculated in conjunction with flow velocity, residence time, and other parameters that are specific to the column setup. This information permits the user to determine whether mineral precipitation is significant in terms of the long-term performance of the permeable barrier.

CHAPTER 4

TREATABILITY TESTING

Following site characterization and identification of prospective reactive media candidates, bench-scale treatability testing is conducted to aid the design. Treatability testing serves the following purposes:

- Screening and selecting a suitable medium (e.g., iron) for the reactive cell
- Estimating the half-life of the degradation reaction and determining flowthrough thickness of the reactive cell
- Evaluating the longevity of the wall.

Treatability tests can be conducted in a batch or column (continuous) mode. Most researchers now agree that batch tests are useful mainly as an initial screening tool for evaluating different media or for assessing the degradability of contaminants hitherto known to be recalcitrant (Sivavec, 1996; ETI, 1996). For most other purposes, researchers favor column tests for the following reasons:

- Design parameters are determined under dynamic flow conditions. As concentrations of contaminants and inorganics change with the distance traveled through the reactive cell, they can be measured by installing a number of intermediate sampling ports along the length of the column.
- Half-lives measured through column tests are generally more reliable than half-lives measured through batch tests.
- Nonlinear sorption to nonreactive sorption sites (Burris et al., 1995) is better simulated in columns.
- Reaction products may accumulate in the batch reactor, whereas they may be washed away in columns.

Various types of water may be used to run treatability tests:

- Deionized water spiked with the targeted contaminant(s)
- Clean groundwater from the site spiked with the desired concentration of chlorinated solvents
- Contaminated groundwater from the site.

Some researchers use clean, rather than contaminated, water in treatability tests; screening of new reactive media may be conducted with clean deionized water, whereas other treatability tests may be conducted with uncontaminated groundwater from the site. The uncontaminated water is spiked with known concentrations of the contaminants. In this way, researchers can better control or change feed concentrations. It is also easier to collect and ship uncontaminated groundwater from the site than it is to collect and ship groundwater already contaminated with volatile organics. It is important to run at least some tests with groundwater from the site (clean or contaminated) because of the important role played by native inorganic parameters in the site groundwater.

4.1 BATCH TESTING

Batch experiments generally are conducted by placing the media and contaminant-spiked water in septum-capped vials with no headspace. When samples are drawn from the vial for analysis, the vial is either sacrificed or nitrogen is added to fill up the headspace created (Sivavec, 1996). Nitrogen can be introduced into the vial by sampling with the dual-syringe technique. As the sample is drawn into one syringe, the other syringe (filled with nitrogen) slowly releases nitrogen into the headspace. Alternatively, deionized water may be used to replace the liquid withdrawn for analysis. In this way, concentrations of organics can be measured as a function of time over multiple sampling events. The formation of reaction by-products (hydrocarbon gases) in the vial can also be measured to differentiate any loss of reactants due to sorption rather than reaction.

Batch tests are useful screening tools because they can be run quickly and inexpensively. However, care should be taken in extrapolating the results to dynamic flow conditions. For example, O'Hannesin (1993) found that the column half-lives for TCE and PCE exceeded batch values by factors of 3 and 2, respectively, even though a higher iron-to-solution ratio was used in the columns than in the batch tests.

4.2 COLUMN TESTING

The main objective of column tests is to estimate the half-life of the degradation reaction. The half-lives of the organic contaminants and their byproducts are then used to either select the reactive medium or design an appropriate flow-through thickness for the reactive cell.

4.2.1 Design and Implementation of Column Tests

The design of a typical column setup is shown in Figure 4-1. A single column with multiple sampling ports along its length is used. The column is made from glass or Plexiglass™. Strictly speaking, glass should be expected to have the least adsorptive or reactive effect with chlorinated organic compounds; however, no significant

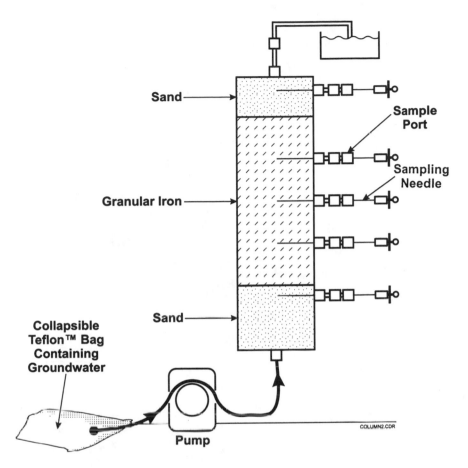

FIGURE 4-1. Typical column setup

loss of organics has been found by researchers using Plexiglass™ columns. All fittings are stainless steel. Tubing is either stainless steel or Teflon™. A small section of tubing through the peristaltic pump is made of Viton™ for added flexibility.

The column is packed with the reactive medium in such a way as to ensure a homogeneous matrix. One way of doing this is to make small aliquots of well-mixed media (e.g., iron and sand) and fill the column in small batches with each aliquot. Optionally, a section of sand may be placed above and below the reactive medium in the column to ensure good flow distribution. The average bulk densities, the porosity, and the pore volume can be measured by weight.

The feed water is placed in a collapsible Teflon™ bag to prevent headspace as the bag empties out. The bag is filled by gravity flow to avoid aeration of the water. Water is circulated in the column from bottom to top to better simulate the flowrates likely to be encountered in the field. Sampling ports are equipped with gastight and watertight fittings. A nylon swage lock fitting may be used or a septum may be

crimped onto the sample port. It is best to leave the sampling syringe needles per-
manently inserted into the column, with the tip at the center of the column. Valves
with luer-lock adapters are attached to the protruding ends of the needles outside the
column (Sivavec, 1996; ETI, 1996). A luer-lock plug is used to seal the needle
between samples. Figure 4-2 shows a typical column test in progress.

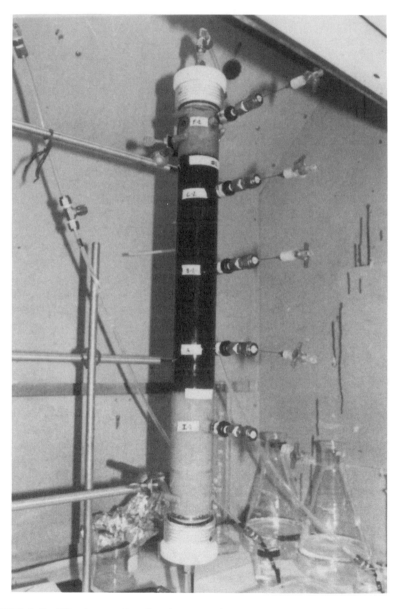

FIGURE 4-2. Photograph of column setup

Sampling should begin only after the concentration distribution in the column has reached steady state; that is, the net contaminant mass entering the column should be equal to the mass degraded within the column. It is generally necessary to run several pore volumes of contaminated water through the column before it reaches steady state, according to Burris et al. (1995). They also showed that the time (pore volumes) required to reach steady state varies with contaminant type. For example, water contaminated with PCE requires a longer time to reach steady state than does water contaminated with TCE.

Whenever a sample is to be drawn, a syringe is attached to the adapter on the needle and the sample is collected after a small amount of water is purged from the needle. The sample is drawn very slowly to create minimum disturbance in the flow. Most researchers conduct column experiments at room temperature. It is important to note, however, that temperature may be an important factor influencing reaction rate.

The flowrate through the columns can be set to approximate site conditions. Therefore, it is desirable to have good groundwater velocity data from the site at the location of the proposed barrier. We also need to take into account the fact that local groundwater velocity through the permeable barrier may be considerably higher than in the surrounding aquifer, especially in a funnel-and-gate system. Section 5.1 describes how expected groundwater velocity through the reactive cell can be determined through particle tracking maps. Actual flowrates through the column can be measured by collecting a timed volume of effluent. The experiment could be repeated over a range of flowrates to account for seasonal variations and other uncertainties. Interestingly, flowrate may not be a critical parameter for column testing. Gillham and O'Hannesin (1992) found that degradation rates were insensitive to flowrates in the range tested (59 to 242 cm/day). However, once degradation rates have been determined through column tests, designing the flowthrough thickness of the reactive cell does require an accurate estimate of groundwater velocity.

Concentration profiles may be generated periodically for the distribution of chlorinated organics in the column by collecting and analyzing samples from the influent, the effluent, and the intermediate sample ports after every 5 to 10 pore volumes. Eh and pH profiles of the column may be generated less frequently because of the higher sample volumes required for taking these measurements with typical probes. The column influent and effluent should be analyzed also for inorganics, such as major cations (Ca, Mg, Na, Fe, Mn, and K), major anions (Cl, SO_4, NO_3, and NO_2), and alkalinity (bicarbonate).

Analysis of water samples collected from the column is done by the same general methods used for analyzing groundwater samples during site characterization (see Sections 2.2 and 2.3). Concentrations of chlorinated VOCs can be measured using a gas chromatograph-flame ionization detector (GC-FID) with purge and trap equipment. Water samples typically are drawn through sampling needles into a gastight syringe and are injected directly into the purge and trap through luer-lock adapters. Although chlorinated compounds can be detected using an electron capture detector (ECD), the GC-FID is suitable for general-purpose work because it can detect both a broad range of low-molecular-weight chlorinated compounds (e.g., TCE, DCE, and VC), as well as nonchlorinated hydrocarbon byproducts such as ethene or ethane. Normally, the instrument is calibrated to detect compounds at the lowest concentrations feasible. A typical detection limit for chlorinated hydrocarbons is 2 µg/L, provided that there is no strong matrix interference that requires dilution of the primary sample.

Anions typically are measured using ion chromatography (IC) and cations by inductively coupled plasma (ICP). Detection limits for inorganic constituents also can depend on the matrix. There should be very little problem with analyte interference when deionized or low-TDS water (synthetic or actual groundwater) is used; however, this may not be the case when high-TDS water is used.

Eh and pH are measured using appropriate probes (usually combination electrodes) in water samples immediately after they are withdrawn from the column. Accurate pH measurements can be taken only when the water is buffered or contains adequate concentrations of strong acid or strong base. Because most waters are near neutral to slightly alkaline and metallic compounds may raise the pH above 9, the pH range of 6 to 8 may be the most difficult range in which to obtain accurate readings. This is particularly true when the water in the column contains no buffer, such as carbonate.

Similarly, accurate Eh readings taken with a platinum electrode cannot be obtained unless the system is buffered with respect to electron transfer reactions; such a system is referred to as being "poised." When a system is not well poised, Eh measurements do not reflect the abundance of electrons due to the combinations of half-cell couples. This can happen when the species present are not electroactive, i.e., they do not react rapidly on the electrode surface. Examples of nonelectroactive species are sulfate (SO_4^{2-}), bicarbonate (HCO_3^-), methane (CH_4), and N_2. Complementary redox species that are electroactive include bisulfide (HS^-) and ammonium (NH_4^+). The problem of an unpoised system may exist only in water samples collected outside the reactive media. Within the reactive media the system should be well poised, for example, due to the $Fe^0/Fe(II)$ redox couple and combinations of others couples. When the accuracy of Eh measurements is in question (because the system is poorly poised), Eh may be calculated using ion concentrations. Some of the geochemical equations mentioned in Appendix C, Section C.1 permit the user to calculate Eh based on HS^-/SO_4^{2-}, N_2/NH_3^+, CH_4/HCO_3^-, or other redox couples.

Dissolved oxygen is difficult to measure off line, and may require a flowthrough cell or an in-line probe that excludes atmospheric oxygen from the sample. The dissolved oxygen concentration normally will be negligible when redox potential is negative, as should be the case when highly reducing metals such as iron or zinc are in equilibrium with the water. Therefore, dissolved oxygen measurements could be omitted during column tests, particularly if Eh can be measured with confidence.

4.2.2 Interpreting Column Data

For each water flow velocity and each column profile, chlorinated VOC concentrations are plotted initially as a function of distance through the reactive column (ETI, 1996). When the flowrate and porosity are known, distances through the column can be converted easily to residence times. A graph of VOC concentrations (μg/L) versus residence time (in hours) can then be generated (see an example plot in Figure 4-3). If C_o is the initial concentration of the chlorinated VOC contaminant and C is its concentration after time t, then a degradation rate constant, k, can be calculated for each concentration profile using first-order kinetics.

$$C = C_o \, e^{-kt} \qquad \text{Eqn. 4-1}$$

When ln (C/C_o) is plotted against time in hours (see an example plot in Figure 4-4), the slope of the fitted line is the reaction rate, k (hr^{-1}). The degree of fit can be determined by calculating the correlation coefficient (r^2). The r^2 value indicates how well the pseudo first-order model fits the experimental data. Once the rate constant is known, a half-life ($t_{1/2}$) can be estimated for each organic contaminant of interest (TCE, PCE, etc.) in the influent. A half-life is the time period required to reduce the concentration of a contaminant by half. Table 9-1 in Chapter 9 shows the estimated half-lives for various chlorinated VOCs.

$$t_{1/2} = \frac{\ln(2)}{k} = \frac{0.693}{k} \qquad \text{Eqn. 4-2}$$

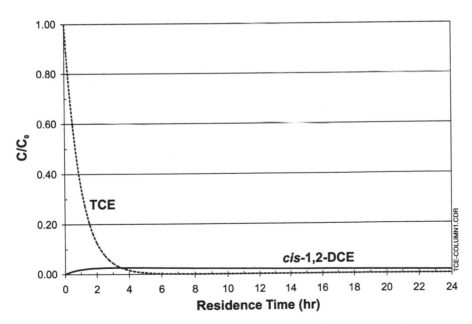

FIGURE 4-3. Column concentration profile of TCE and one of its byproducts, *cis*-DCE

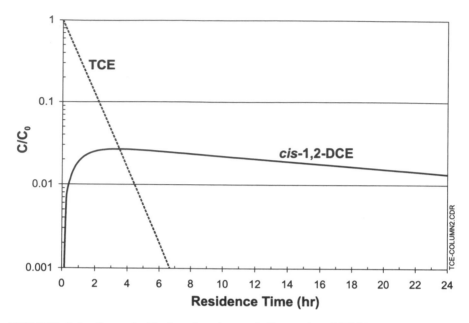

FIGURE 4-4. Psuedo first-order degradation rate of TCE

When comparing the half-lives obtained for the same compound in columns with different reactive media, the reactive medium which provides the shortest half-life generally is selected. Costs, availability, buffering effects, and other factors also may be considered, as described in Chapter 3.

The required *residence time* for the permeable barrier design can be determined in several ways. Residence time can be estimated simply by the number of half-lives required to bring the concentration of the chlorinated VOC down to its MCL. For example, if TCE enters the reactive cell at 1,000 μg/L, eight half-lives are required to degrade TCE to an MCL of 5 μg/L. If the half-life of TCE from the column test was determined to be 2 hours, the required residence time in the reactive cell would be at least 16 hours. If there is more than one VOC of interest in the influent, the residence time is determined from the VOC with the longest half-life.

Alternatively, a residence time, t_w, can be estimated by rearranging Equation 4-1, as:

$$t_w = (1/k) \ln (C_o/C) \qquad \text{Eqn. 4-3}$$

However, this type of estimation does not take into account the residence times required to degrade byproducts (DCE, VC) that may be produced in the reactive cell. It may be that these byproducts are being produced and degraded concurrently, making it difficult to determine the reaction rates for the byproducts.

The following procedure for determining residence time may provide a better estimate (ETI, 1996). The concentrations of all the chlorinated VOCs (influent and byproducts) are plotted against residence time in the column on the same graph. Figure 4-5 is an example of such a graph. The required residence time is the longest time required to bring all the contaminants down to their MCLs. In the example (Figure 4-5), the residence time is driven by the time required to bring VC down to its MCL. This type of estimation may require testing in longer columns or multiple columns in series if the simulated groundwater velocity is high.

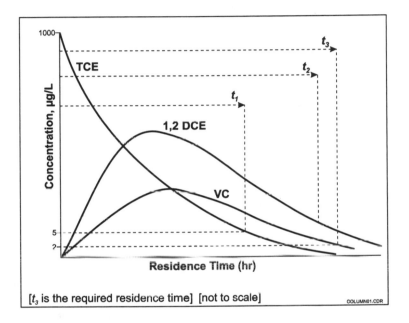

FIGURE 4-5. Example of a column profile of VOC concentrations

The parameters pH and Eh can also be plotted against residence time (distance through column). Typically, as conditions become more anaerobic in the column, Eh should decline and pH should increase with increasing distance. Changes in anions, cations, and alkalinity between influent and effluent are also examined to understand the geochemical behavior of the system. Loss of dissolved calcium or magnesium could indicate the potential for precipitate formation in the reactive cell. Increase in dissolved iron in the effluent could indicate losses of ferrous iron from the cell. A geochemical evaluation of the inorganic chemistry data from column tests can provide a good basis for reactive media selection and longevity assessment (see Section 5.2).

4.2.3 Safety Factors

Some safety factors must be considered in adjusting the degradation rate from laboratory data to field application. One important adjustment is the temperature. The temperature of the groundwater in the field application (typically 10 to 15°C) is generally lower than the room temperature of the experiment (typically 20 to 25°C). The empirical residence time may need to be increased to account for the lower temperature. For example, Senzaki and Kumangai (1988a and b) found that the half-life of 1,1,2,2-tetrachloroethane increased by 10 percent when temperature declined from 20 to 10°C. Jeffers et al. (1989) provide a discussion of the use of Arrhenius temperature dependence to adjust for the effects of temperature on degradation rate of organic compounds. The Arrhenius equation relates the reaction rate (k) to absolute temperature (T) as follows:

$$k \propto e^{-E/RT}$$ Eqn. 4-4

where E is the activation energy, and R is the universal gas constant (8.314 Joules/mol kelvin). Equation 4-4 can be rearranged as:

$$\ln k = (\ln A) - (E/RT)$$ Eqn. 4-5

where k = first-order reaction rate constant
       A = frequency factor for the reaction
       E = activation energy
       R = ideal gas constant
       T = absolute temperature

A plot of ln k versus 1/T should give a straight line with a slope of −E/R and an intercept on the 1/T axis of ln [(A)/(E/R)]. Experimental data from controlled-temperature column tests (ETI, 1997) resulted in the linear plot of the TCE degradation rate constant versus temperature in Figure 4-6. The plot indicates that, at 15°C in the field, TCE degradation rates could be expected to decline by a factor of 1.4 from those measured in the laboratory at 23°C. Field observations at a test site in New Jersey have shown that the degradation rate declines by a factor of 2 to 2.5 at temperatures of 8 to 10°C compared with laboratory rates. Similar results have been observed at other field sites.

Temperature versus reaction rate relationships have not as yet been determined experimentally for PCE. Given PCE's similar behavior to TCE in dehalogenation reactions, it may be assumed that a similar temperature factor would apply. During the design phase at existing sites, no temperature factors were applied to DCE or VC degradation rates, as measured rates of DCE and VC degradation in controlled-temperature experiments to date have shown little influence of temperature.

The bulk density of the reactive cell in the field is generally lower than the bulk density measured in the laboratory because of different settling conditions for

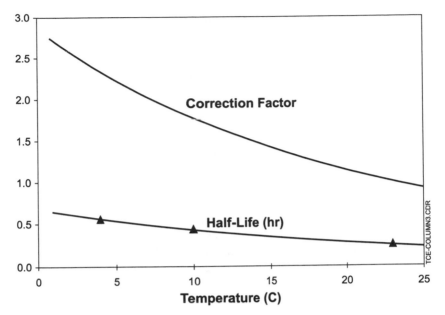

Reprinted with permission from Envirometal Technologies, Inc., Guelph, Ontario (ETI, 1997).

FIGURE 4-6. Correlation of TCE degradation rates with temperature

the medium (ETI, 1996). Therefore, the surface area of reactive medium per unit volume of groundwater in the field may be lower than that in columns. Degradation rates (or half-lives) are proportional to the specific surface area of the reactive medium (Gillham, 1996; Sivavec and Horney, 1995). The field residence time must be increased to account for the lower expected reactive surface area to volume of solution ratio. Currently, there is no clear indication of how large the bulk density correction factor should be. To some extent, this factor would depend on the efficiency of the emplacement and how well the reactive medium consolidates after emplacement. Gillham et al. (1993) reported that an increase in the surface area of iron by a factor of 5 caused the half-life for TCE to decline by a factor of about 2.5. Reduced iron surface area per unit volume of groundwater is the reason why 100 percent iron degrades faster than iron-sand mixtures. Also, finer iron granules generally have larger surface areas and faster degradation rates.

Recent reports from field installations indicate that a bulk density of approximately 160 lb/ft^3 is generally obtained with granular iron. This would indicate a field porosity of 0.65 instead of 0.45, as reported from laboratory column studies. This suggests that a bulk density correction factor of 1.5 needs to be applied to the reaction rate or half-life determined from laboratory column tests.

Additional safety factors may be applied, depending on the degree of uncertainty in key data, such as groundwater velocity and influent contaminant concentrations.

4.2.4 Determining Flowthrough Thickness of the Reactive Cell

Based on the groundwater velocity expected in the field reactive cell and the required residence time, the flowthrough thickness (b) of the field reactive cell can now be determined as:

$$b = V_x \bullet t_w \hspace{4cm} \text{Eqn. 4-6}$$

It should be noted that the groundwater velocity through the reactive cell is usually higher than the groundwater velocity in the aquifer because the conductivity of the reactive cell is usually higher than that of the aquifer. Hydrogeologic modeling is used to determine the expected groundwater velocity through the reactive cell. Additional safety factors may be incorporated into the calculated thickness to account for seasonal variations in the flow, potential loss of reactivity of the iron over time, and any other field uncertainties. Section 5.1 includes a discussion on hydrogeologic and geochemical factors affecting flowthrough thickness.

4.3 ACCELERATED AND LONG-TERM COLUMN TESTING

Recently, some researchers have tried to use accelerated column tests to estimate the longevity of the permeable barrier (Sivavec, 1996). Their main approach has been to use accelerated aging techniques. Initially, degradation rates are measured in a column at the expected local velocity in the reactive cell. The flow is then increased to "age" the iron by passing a large number of flow volumes. Periods of low and high flow are alternated. At each low flow step, as soon as steady state is reached, measurements are conducted to estimate reaction rates, porosity losses (measured through tracer tests), inorganic profiles, and reaction products.

Caution should be exercised in interpreting the results of accelerated column tests. Equating 100 pore volumes at 20 feet/day in the laboratory with 1,000 pore volumes at 2 feet/day in the field may not provide an exact estimate, because the lower residence time in the accelerated column test may underestimate the amount of precipitation. Therefore, extrapolation to the field may not be easy. The advantage, on the other hand, is that a large number of pore volumes can be passed through the reactive medium in a short time. An alternative to the low-flow/high-flow approach is to run the columns at field groundwater velocities for very long times, i.e., for several months or years. An important objective in all such long-term tests is to study whether the inorganic profile ever levels off (Sivavec, 1996). Another objective is to ensure that the selected reactive medium can sustain its reactivity over longer periods of time (over several pore volumes). There is an indication that some reactive media (e.g., Fe-Ni bimetallic) may provide high reaction rates initially, but these rates may decline over long periods of time (Sivavec et al., 1997).

To extrapolate from column tests to field lifetime, the number of pore volumes is used as a scaling factor. For example, to translate from pore volumes to years, researchers may assume a velocity of 1 foot/day and a 4-foot wall thickness. With these conditions, 182 pore volumes would be equivalent to a lifetime of 2 years. Researchers suggest following the lifetime of each section of the column separately, because fouling is more likely to occur near the influent end where oxygen enhances precipitation (Sivavec, 1996).

4.4 ESTIMATING THE PERMEABILITY OF THE SELECTED REACTIVE MEDIUM

Hydraulic conductivity (K) of the reactive medium is required to determine the flow velocity and residence time of groundwater through the reactive cell. Ideally, the K value for unconsolidated media is determined from constant head permeameter tests (Fetter, 1994). These tests are most reliably conducted in laboratories with conventional permeameter facilities. However, it may also be possible to use the laboratory treatability test columns for estimation of K by setting up the columns as constant head permeameters.

Constant head permeameters consist of an inlet tube with water level (head) maintained at a height slightly above the outlet level of the column. The water is allowed to flow through the reactive medium in the column until steady-state flow is obtained and the volume of water flowing out over a period of time is measured. K is determined from a variation of Darcy's law:

$$K = \frac{V \cdot L}{A \cdot t \cdot h}$$

Eqn. 4-7

where V is the volume of water discharging in time t, L is the length of the reactive medium sample, A is a cross-sectional area of the sample, and h is the hydraulic head difference across the column. It is important to prepare a uniformly packed column, maintain a hydraulic gradient across the column similar to that expected in the field, and ensure the absence of air bubbles in the column.

CHAPTER 5

MODELING TO SUPPORT THE PERMEABLE BARRIER DESIGN

Modeling enables an understanding of the implications of site characterization information and treatability data. Hydrogeologic modeling is conducted for the following reasons:

- Determine an approximate location and configuration for the permeable barrier with respect to the groundwater flow and plume movement.

- Estimate the expected groundwater flow velocity through the reactive cell.

- Determine the width of the reactive cell and, for a funnel-and-gate configuration, the width of the funnel.

- Estimate the hydraulic capture zone of the permeable barrier.

- Determine appropriate locations for performance and compliance monitoring points (discussed in Chapter 7).

- Evaluate the hydraulic effects of potential losses in porosity (and potential for flow bypass) over the long term.

- Evaluate the potential for underflow, overflow, or flow across aquifers.

- Incorporate the effects of shifts in groundwater flow direction into the design.

- Incorporate site-specific features such as property boundaries, building foundations, buried utilities, etc., into the design.

Appendix B, Section B.1 describes the hydrogeologic computer codes that are available to support permeable barrier design. For most practical purposes, commercially available models such as MODFLOW (flow model) and its enhancements are sufficient for the design, although comparable commercial or proprietary models may be used as well.

Geochemical models available for evaluating permeable barriers are described in Appendix C, Section C.1. Commercially available models such as PHREEQ are generally sufficient for the purpose. Most available models are equilibrium models, in which reaction kinetics are not incorporated. However, these models can play an important role in understanding the potential for various reactions as the inorganic parameters in the groundwater (e.g., SO_4^{2-}, Ca^{2+}, etc.) pass through the reactive cell. This relates to issues of media selection and longevity of the barrier.

5.1 HYDROGEOLOGIC MODELING APPROACH FOR DESIGN AND MONITORING OF PERMEABLE BARRIERS

Hydrogeologic modeling can be used at several stages of the permeable barrier technology implementation. This includes the initial feasibility assessment, the site selection, design optimization, design of performance monitoring network, and longevity predictions. The major advantage of constructing a detailed ground-water flow model is that several design configurations, site parameters, and performance and longevity scenarios can be readily evaluated once the initial model has been set up. Thus the combined effect of several critical parameters can be incorporated simultaneously into one model. Groundwater modeling has been used at most previous permeable barrier installations. In most cases, groundwater flow models have been used in conjunction with particle tracking (solute transport) codes to construct flownets showing travel paths and residence times through the reactive cell. The models are usually set up after laboratory column tests have shown the feasibility of the degradation, and the reaction half-lives and the resulting residence time requirements have been determined.

This section describes the use of models in the evaluation of permeable barrier design and performance. The general requirements of the modeling codes useful for permeable barrier application, a brief overview of the modeling methodology, descriptions of the available codes, and a review of previous modeling studies for permeable barrier design are presented in Appendix B, Section B.1.

The two primary interdependent parameters of concern when designing a permeable barrier are *hydraulic capture zone* width and *residence time*. Capture zone width refers to the width of the zone of groundwater that will pass through the reactive cell or gate (in the case of funnel-and-gate configurations) rather than pass around the ends of the barrier or beneath it. Capture zone width can be maximized by maximizing the discharge (groundwater flow volume) through the reactive cell or gate. Residence time refers to the amount of time contaminated groundwater is in contact with the reactive medium within the gate. Residence times can be maximized either by minimizing the discharge through the reactive cell or by increasing the flowthrough thickness of the reactive cell. Thus, the design of permeable barriers must balance capture zone width (and discharge) against the required residence time. Contamination passing outside the capture zone will not flow through the reactive cell, and so will not be treated. Similarly, if the residence time in the reactive cell is too short, contaminant levels may not be reduced sufficiently to meet regulatory requirements.

A number of numerical simulations were performed (Battelle, 1996) to illustrate the design and evaluation of permeable barriers through modeling. The methodology and results of this modeling are described in Appendix B, Section B.2 and summarized below.

5.1.1 Modeling Approach for Relatively Homogeneous Aquifers

An illustration of the modeling approach for relatively homogeneous aquifers is presented in Section B.2 (Battelle, 1996). A relatively homogeneous aquifer can be modeled using two-dimensional (2D) versions of flow and particle-tracking codes. At most existing permeable barrier application sites so far, this simplified approach has been used to locate and design the barrier. Permeable barrier features, such as the reactive cell, pea gravel, or funnel walls, can be inserted into the baseline aquifer model as heterogeneities with the appropriate hydraulic conductivities. The hydraulic conductivity of the reactive cell can be estimated based on the particle size of the reactive medium used or, for more certainty, measured through laboratory permeability testing. Design parameters, such as hydraulic capture zone width (feet), residence time within the reactive cell (hours or days), and groundwater discharge through the gate (ft^3/day) can then be estimated for each simulation. The hydraulic capture zone width in each simulation can be determined by tracking particles forward through the gate.

Figure 5-1 shows the particle tracking results for a permeable barrier in a relatively homogeneous aquifer. The hydraulic capture zone is symmetrical and extends beyond the width of the gate. Remixing of the water flowing through the gate and water flowing around the barrier takes place downgradient. This type of modeling is a useful tool for designing the dimensions of the reactive cell (gate) and funnel, as well as simulating scenarios for different configurations. For example, the funnel walls could be eliminated in one simulation or the width of the reactive cell (gate) could be increased in another simulation.

Particle-tracking codes also can be used to design a performance-monitoring network along specific flowpaths for evaluation of potential contaminant breakthrough or bypass. This approach is especially useful if tracer tests are to be used to evaluate permeable barrier performance.

5.1.2 Modeling Approach for Heterogeneous Aquifers

Modeling studies and barrier design at most existing permeable barrier sites so far have been primarily based on the assumption that the aquifer sediments in the vicinity of the permeable barrier are homogeneous. However, at many sites, there may be strong heterogeneity in the sediments. This heterogeneity is mainly due to the variations in depositional environments of the sediments. The general implications of heterogeneity are that more detailed site characterization is required and more complex models are needed. The symmetrical capture zones seen in homogeneous sediments become asymmetrical and difficult to predict without detailed characterization and modeling.

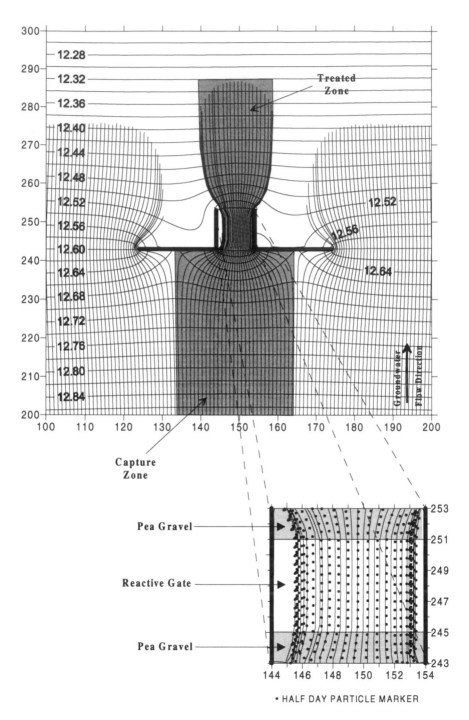

FIGURE 5-1. Simulated particle pathlines showing capture zone

Figure 5-2 shows the results of modeling conducted at a permeable barrier site in California (Battelle, 1996; and PRC, 1996). The capture zones at this site, as seen from these particle tracking maps, are highly *asymmetrical*. In the less-permeable layers (Layers 1 and 2), there is hardly any movement of particles over 25 days. In the more-permeable Layer 3, the particle movement is very fast upgradient of the gate but very slow upgradient of the funnel walls. In the more-permeable Layer 4, the particle movement is very fast in front of the west funnel wall but somewhat slower on the east side. These irregularities exist because the lower part of the permeable barrier (Layers 3 and 4) is in a high-conductivity sand channel, whereas the upper part is located in lower-conductivity interchannel deposits. The location of the sand channels at the site was determined from existing base-wide site characterization maps and from localized CPT data generated during additional site characterization activities conducted to aid the design of the barrier. The irregularities in flow may result in vastly different residence times in the reactive cell. Pea gravel sections along the upgradient and downgradient edges of the reactive cell help to homogenize the flow vertically and horizontally to some extent.

A similar situation is reported by Puls et al. (1995) for the Elizabeth City, North Carolina site. At this site the geology is characterized by complex and variable sequences of surficial sands, silts, and clays. Groundwater flow velocity is extremely variable with depth, with a highly conductive layer at roughly 12 to 20 feet below ground surface. The reactive cell was emplaced in this sand channel (see Figure B-7 in Appendix B, Section B.2).

These examples illustrate the need for placing the reactive cell in a zone of high conductivity that forms a preferential pathway for most of the flow and contaminant transport through the aquifer. Additionally, the dependence of capture zones on aquifer heterogeneities illustrates the need for detailed site characterization and adequate hydrogeologic modeling prior to permeable barrier design and emplacement. Particle tracking simulations, such as the one shown in Figure 5-2, along with a flow model based on good site characterization, can also help in optimizing monitoring well locations for evaluating the performance of the barrier.

5.1.3 Modeling Different Permeable Barrier Configurations and Dimensions

One of the advantages of groundwater modeling is that numerous different configurations of the permeable barrier systems can be simulated to select the design most suitable for a site. Thus relative benefits and limitations of various options may be evaluated prior to expensive field installations. Simulation of one of the common designs, the simple funnel-and-gate system, has already been presented in this chapter and in Section B.2. In addition, another simulation incorporating a funnel-and-gate design in heterogeneous media is shown in Figure 5-2. In this section, two more design scenarios are presented to illustrate the use of groundwater flow models in designing and optimizing permeable barriers (Battelle, 1996).

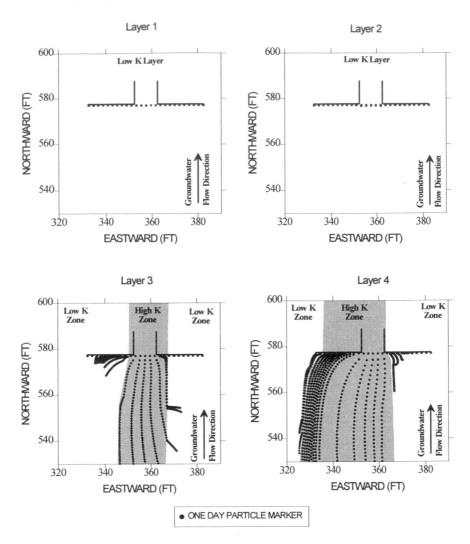

FIGURE 5-2. MFA funnel-and-gate backward particle tracking showing the effect of heterogeneity on capture zones (Battelle, 1996)

An example simulation of the first scenario with a continuous barrier is shown in Figure 5-3. This simulation consists of a 10-foot-long section of reactive media having a 6-foot thickness in the direction of flow. The aquifer is simulated as a single layer having uniform hydraulic properties with a conductivity of 10 feet per day. The reactive media are simulated with a hydraulic conductivity of 283 feet per day. The flow field was simulated with a gradient of 0.005 foot/foot. Particle tracking techniques were used to delineate the capture zone of the reactive media by delineating flowpaths for 180 days. As indicated by the dashed lines, the capture

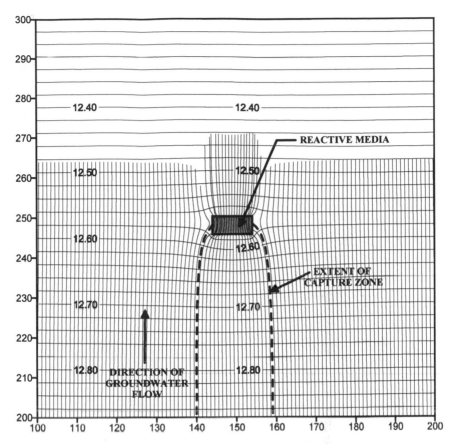

FIGURE 5-3. Simulated capture zone for a continuous barrier scenario showing flowpaths for 180 days

zone has a width greater than the 10-foot length of the reactive media. The width of the capture zone will increase or decrease as the ratio of the reactive media hydraulic conductivity to the aquifer hydraulic conductivity increases or decreases, respectively. Residence time through the reactive media can be estimated using particle tracking methods to ensure sufficient residence time for the degradation reactions to occur. In this case, where no funnel walls are used, several short flowpaths into and out of each end of the reactive media occur. Groundwater flowing along these paths does not pass through the entire thickness of the reactive media, and therefore, entrained contaminants may not be fully degraded in these instances unless appropriate safety factors are incorporated into the design.

The second scenario involves simulation of a permeable barrier with two 8-foot-diameter caissons containing reactive media installed in a funnel-and-gate type configuration similar to that shown in Figure 1-1d. Slurry walls were selected as the materials to be used to construct the funnel walls. A 4-foot by 4-foot zone within the caissons represents the area for the reactive media emplacement. Both up- and downgradient pea gravel are simulated for the areas immediately adjacent to the reactive media. The slurry walls extend 89 feet between the two gate locations and 33 feet on each end of the installation. A single layer, 2D groundwater flow model was used, with the assumption that the aquifer had a uniform hydraulic conductivity of 8.5 ft/day. A K of 283 ft/day was assigned to the reactive media; pea gravel K = 2,830 ft/day, and slurry walls K = 2×10^{-6} ft/day. A gradient of 0.002 was imposed on the flow system, resulting in a groundwater flow velocity of about 0.05 ft/day. The calculated flow field was used to estimate the capture zones (Figure 5-4) for the

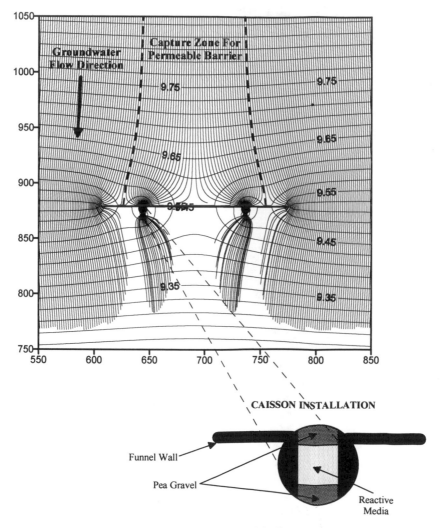

FIGURE 5-4. Capture zone for a permeable barrier with two caissons and funnel walls. Flowpaths for 5,000 days shown

simulated caisson-and-gate installation. Particle tracking techniques using RWLK3D were employed to delineate the flowpaths for 5,000 days. As seen in Figure 5-4, the capture zones for this homogeneous system are roughly symmetrical with flow divides at the approximate midpoints of the funnel walls. However, in heterogeneous systems these capture zones are likely to be asymmetrical. Another noteworthy aspect of this illustration is that it takes about 10 years for 200 feet of plume capture. This indicates that at low groundwater velocity sites, the remediation of large plumes can take a very long time. At sites such as this, it may be useful to consider using recharge wells upgradient of the permeable barrier to increase hydraulic gradient to accelerate the plume capture.

Models similar to those shown in this section can be set up to evaluate several designs under site-specific conditions before selection of an appropriate design for the final installation. It should be noted that the scenarios presented above are based on highly simplified and idealized situations. For models based on field conditions, it is critical to incorporate heterogeneities, seasonal variations in flow conditions, potential vertical flow gradients, site features such as buried utilities and buildings, and calibration into the model for realistic prediction.

Interrelated with the modeling of different permeable barrier configurations is the modeling of different *dimensions* of the barrier for a given configuration. Different widths of a continuous reactive barrier, gate, or funnel can be simulated to evaluate any trade-offs that may occur between various design parameters (e.g., hydraulic capture zone width versus residence time in the reactive cell). For example, the illustrative modeling scenarios in Section B.2 of Appendix B provide the following considerations for permeable barrier design:

- While designing the dimensions of the reactive cell, it is important to note that aquifer hydraulic conductivity ($K_{aquifer}$) is the sensitive parameter for discharge and residence time through the reactive cell when the ratio of reactive cell hydraulic conductivity (K_{cell}) to $K_{aquifer}$ is greater than 5 to 1. Reductions in K_{cell} do not significantly impact discharge and residence times through the gate until the ratio of K_{cell} to $K_{aquifer}$ drops below 5:1, where K_{cell} becomes an increasingly sensitive parameter. This type of analysis can be used with site-specific models to evaluate the effect on the performance of the permeable cell of decreasing reactive cell permeability over time. Appropriate safety factors (in terms of additional reactive cell width or larger particle size reactive medium) then can be incorporated into the design for anticipated changes in capture zone and residence time.

- As discharge through the reactive cell increases, capture zone width increases, and travel time through the reactive cell (residence time) decreases. For the scenarios simulated in this report, residence times in the reactive cell ranged from over 200 days for low-K (0.5 foot/ day) aquifers to roughly 1 day for higher K aquifers (100 feet/day). The particle tracking estimates of residence times can be used to

optimize the flowthrough thickness of the reactive cell for achieving
the desired reduction in contaminant levels.

- Particle tracking also may be used to design a performance monitoring
  network along specific flowpaths. This approach is especially useful if
  tracer tests are to be conducted in the reactive cell or its vicinity. Some
  particle tracking codes, such as RWLK3D, can also incorporate the
  solute transport processes. These may be used to evaluate the effects of
  dispersion within the reactive cell. The fastest travel times determined
  from the advective-dispersive simulations would then be used to determine the safety factor required in designing the reactive cell.

- For funnel-and-gate configurations, hydraulic capture zone width
  appears to be most sensitive to funnel length and aquifer heterogeneity. Capture zone width is generally greater for higher values of
  K_{cell} when $K_{aquifer}$ is held constant. At ratios greater than 5:1 between
  K_{cell} and $K_{aquifer}$, capture zone width did not change significantly
  when only the K_{cell} was varied. Higher conductivity aquifers have
  larger capture zones relative to less conductive aquifers for the same
  K_{cell}. Capture zone width is more sensitive to variability in $K_{aquifer}$
  relative to changes in K_{cell}.

Similarly, the following design considerations were obtained from the previous modeling studies (Starr and Cherry, 1994; Shikaze, 1996) described in Appendix B, Section B.1:

- In a funnel-and-gate configuration, the maximum absolute discharge
  (groundwater flow volume) through the gate occurs when the funnel
  walls are at an apex angle of 180 degrees (straight barrier).

- For all apex angles, the maximum discharge occurs when the funnel
  is perpendicular to the regional flow gradient.

- A balance between maximizing the hydraulic capture zone size of the
  gate and maximizing the residence time in the reactive cell should be
  achieved through modeling. In general, for a funnel-and-gate system,
  hydraulic capture zone size (or discharge through the gate) and
  residence time are inversely proportional. The residence time can
  usually be increased without affecting the size of the capture zone by
  increasing the width of the gate.

- For funnel walls at 180 degrees (straight barrier), the hydraulic capture zone size (or discharge) increases with increasing funnel width.
  However, the relative capture zone width decreases dramatically as
  the funnel width increases. The relative capture zone width is the
  ratio of the capture zone width to the total width of the funnel-and-
  gate system.

- For a constant funnel width, the absolute and relative capture zone width increases with gate width. Therefore, it is desirable to have a gate as wide as is economical.

- For a given funnel-and-gate design, the capture zone size increases with increase in K_{cell} relative to the $K_{aquifer}$. However, there is relatively little increase in capture zone size when the K_{cell} is more than 10 times higher than $K_{aquifer}$. Therefore, in selecting the particle size of the reactive medium, it is useful to note that the resulting K_{cell} need not be more than about 10 times higher than the K of the surrounding aquifer.

5.2 GEOCHEMICAL EVALUATION FOR PERMEABLE BARRIER DESIGN AND PERFORMANCE

Geochemical modeling is a relatively underutilized area for application to permeable barrier settings. Most designs and performance evaluations have relied more on empirical evidence of reactions between groundwater inorganic parameters (e.g., Ca, Mg, alkalinity) and the reactive medium (iron). One reason for this is possibly the limited availabilities of kinetic geochemical models. Appendix C, Section C.1 describes the various geochemical models available and their potential for application to permeable barrier settings. Section C.2 illustrates how some of these models can be applied to learn about potential reactions and species formation on the reactive cell.

Geochemical modeling is an attempt to interpret or predict the concentrations of dissolved species in groundwater based on assumed chemical reactions. Early efforts were concerned with performing speciation calculations on dissolved inorganic constituents. The models that were developed as an outgrowth of these efforts can be grouped as either forward or inverse models.

In *forward modeling*, reaction progress is governed by thermodynamic expressions, hence the result is an equilibrium prediction. In *inverse modeling*, probable reactions are calculated based on the information supplied at initial and final points along a flowpath, and as such, do not necessarily represent equilibrium. A third type, *reaction-transport modeling*, couples forward modeling with fluid flow and solute transport. This is the newest area of geochemical modeling research, and few highly sophisticated codes have been developed in this area. Most reaction-transport models offer less than three-dimensional (3D) flow fields, have limited capabilities for introducing heterogeneities in the flow regime, and tend to consider only static boundary conditions.

Computer codes typically are used to perform numerical algorithms that model chemical reactivity, hydrochemical transport, or in some cases both. Factors associated with choosing among forward, inverse, and reaction-transport modeling depend on the nature of the geochemical system being considered. Forward modeling may be preferred when only the final outcome of the interaction of groundwater with soils or sediments is desired, i.e., the groundwater composition and mineral saturation index. Inverse modeling provides hydrologic information about an aquifer, such as net mass transfer and mixing, and can be used to determine relative rates of reactions.

Most models allow testing only to see if expected reactions occur to a significant or insignificant extent along a designated flowpath. However, one code (EQ3/EQ6) incorporates reaction rate constants. Another comparison may be made in which forward modeling tests the validity of suspected reactions based on thermodynamic considerations, whereas inverse modeling tests their feasibility based on mass balance considerations. Reaction-transport modeling is distinct, in that it can be used to simulate real transport processes, such as advection and dispersion, in addition to predicting groundwater chemistry. Thus, reaction-transport models may be especially useful for predicting the flowpath of both conservative and nonconservative species.

The availability of data is also a consideration in selecting a geochemical modeling code. Generally, fewer data are needed in forward modeling, whereas fairly complete data are required to achieve definitive results by inverse modeling. As with forward modeling, reaction-transport models may be run with limited chemical data; however, the hydraulic properties of the flow system must be understood.

Not only has it been difficult to use geochemical codes to quantitatively predict the amount of precipitate generated, it is unclear how such a prediction could be correlated to porosity losses and reduced hydraulic conductivity in the reactive cell. Many precipitates could be fine enough to be transported out of the reactive cell through colloidal transport. Therefore, at many permeable sites in the past, a qualitative evaluation of the inorganic data has been used to estimate the potential for precipitate formation in the reactive cell. Site characterization usually gives the first indication. If alkalinity is low or total dissolved solids are high in the groundwater, there is higher potential for precipitates to develop. Selection of appropriate reactive media can alleviate this propensity. After column tests are conducted, changes in dissolved calcium, iron, and alkalinity between column influent and effluent samples can be examined to evaluate the potential for carbonate or hydroxide precipitation.

In general, designers have been satisfied with classifying a site as having either *low* or *high* potential for precipitate formation. An appropriate reactive medium (e.g, iron-pyrite mixture for high-precipitation-potential sites) may then be identified. Greater or lesser safety factors can also be incorporated into the hydraulic design of the permeable barrier according to the type of site. For example, a coarser reactive medium particle size or other method to obtain a higher installed K_{cell} could be incorporated into the design to account for future porosity losses at a high-precipitation-potential site.

CHAPTER 6

EMPLACEMENT TECHNIQUES FOR PERMEABLE BARRIER INSTALLATION

Once the desired location, configuration, and dimensions of the permeable barrier have been determined, a suitable emplacement technique has to be selected. Conventional and innovative techniques that could be used to install a permeable barrier are discussed in detail below and are summarized in Table 6-1. Factors that limit and ultimately decide the type of emplacement method used include the following:

- Depth of emplacement
- Required reactive cell permeability
- Site topography
- Site access and work space
- Geotechnical constraints
- Soil characteristics (of backfill)
- Disposal requirements of contaminated trench spoils
- Costs.

6.1 COMMERCIALLY AVAILABLE TECHNIQUES FOR REACTIVE CELL EMPLACEMENT

The reactive cell is the portion of the aquifer that is modified to contain the reactive medium through which a contaminated plume will flow. Figure 6-1 shows various arrangements of the reactive cell that may be used depending on site-specific hydrogeologic conditions. In a continuous reactive barrier configuration, the reactive cell runs along the entire width of the barrier. In a funnel-and-gate system, only a portion of the total barrier width is taken up by the reactive cell. In some reactive cells, particularly when the surrounding aquifer is heterogeneous, the reactive medium may be bounded on both upgradient and downgradient sides by thinner sections of pea gravel. The pea gravel serves to increase the hydraulic conductivity surrounding the reactive medium and uniformly draw groundwater flow into the reactive cell through a homogeneous material. The pea gravel also provides a homogeneous setting for monitoring the influent to and effluent from the reactive cell.

The reactive cell is generally completed to approximately 2 feet above the water table to allow for water-level fluctuations and consolidation of the reactive

TABLE 6-1. Summary table of various techniques for barrier emplacement

Emplacement Techniques	Maximum Depth (ft)	Vendor-Quoted Cost	Comments
Impermeable Barrier Techniques			
Soil-Bentonite Slurry Wall 　Standard backhoe excavation 　Modified backhoe excavation 　Clamshell excavation	 30 80 150	 $2-8/ft^2 $2-8/ft^2 $6-15/ft^2	Requires a large working area to allow for mixing of backfill. Generates some trench spoil. Relatively inexpensive when a backhoe is used.
Cement-Bentonite Slurry Wall 　Standard backhoe excavation 　Modified backhoe excavation 　Clamshell excavation	 30 80 200	 $4-20/ft^2 $4-20/ft^2 $16-50/ft^2	Generates large quantities of trench spoil. More expensive than other slurry walls.
Composite Slurry Wall HDPE Geomembrane Barrier	100+ 40-50	NA $35/ft^2	Multiple-barrier wall Permeability less than 1×10^{-7}
Steel Sheet Piles Sealable-Joint Piles	60 60	$17-65/ft^2 $15-25/ft^2	No spoils produced Groutable joints
Permeable or Impermeable Barrier Techniques			
Caisson-Based Emplacement	45+	NA	Does not require personnel entry into excavation; relatively inexpensive.
Mandrel-Based Emplacement	190	$7/ft^2	Relatively inexpensive and fast production rate. Multiple void spaces constitute a reactive cell.
Continuous Trenching	35-40	$5-12/ft^2	High production rate. High mobilization cost.
Jetting	200	$40-200/ft^2	Ability to install barrier around existing buried utilities.
Deep Soil Mixing	150	$80-200/yd^3	May not be cost-effective for permeable barriers. Columns are 3 to 5 feet in diameter.
Hydraulic Fracturing	80-120	$2300 per fracture	Can be emplaced at deep sites. Fractures are only up to 3 inches thick.
Jetting Saw Beam	50	$3-4/ft^2	Used for impermeable barriers.
Vibratory Beam	100	$7/ft^2	Driven beam is only 6 inches wide.

HDPE is high-density polyethylene.

medium, although these may vary from site to site. Generally, the reactive cell is keyed in at least 1 foot into the aquitard. In a funnel-and-gate system, the funnel walls are generally keyed in up to 5 feet into the aquitard. If the continuity or integrity of the aquitard is questionable, a geotextile fabric or a concrete floor placed at the base of the reactive cell helps prevent any contamination from entering the reactive cell through underflow. Monitoring well clusters can be installed during

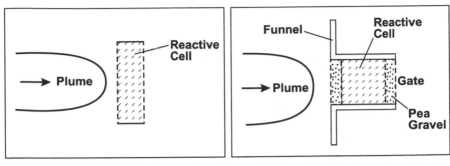

(a) Continuous reactive barrier configuration   (b) Funnel-and-gate configuration

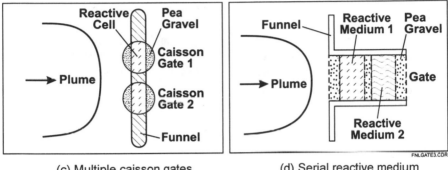

(c) Multiple caisson gates   (d) Serial reactive medium

FIGURE 6-1. Various permeable barrier configurations

cell construction within the reactive medium or in the upgradient and downgradient pea gravel.

Four emplacement techniques are discussed in Sections 6.1.1 through 6.1.4: conventional trench excavation, caisson-based emplacement, mandrel-based emplacement, and continuous trenching. All four have been used at previous sites for reactive cell emplacement. Trench excavation has been widely used. However, there is considerable interest in other techniques, such as caisson emplacement, as deeper plumes are targeted.

6.1.1 Conventional Trench Excavation

Depending on the design of the permeable barrier, installation of the reactive cell may require the excavation of a trench that will house the reactive medium. Backhoes and clamshells are the most common types of equipment used for conventional trench excavation.

To ensure trench wall stabilization during cell construction, several techniques are used. Temporary steel sheet piles can be driven into the ground around the perimeter of the intended reactive cell prior to excavation, then reinforced with bracings. Sheet piling can also be used to temporarily separate the reactive medium and pea gravel sections within the reactive cell (Figure 6-2). Dewatering of the trench may be required if high water tables are present and sheet piling cannot prevent groundwater seepage into the reactive cell. Another option involves excavation under the head of a biopolymer slurry (Owaidat, 1996). The slurry, which is composed of powdered guar bean, acts to maintain the integrity of the trench walls during installation of the cell. The guar gum will later biodegrade to mostly water after wall completion, and will have minimal effect on the permeability of the trench walls. A third method of trench stabilization involves the use of a trench box to create void space during the installation of either impermeable or permeable material (Breaux, 1996). However, the drawbacks to this method are that a trench has to be completely excavated before the box can be installed and temporary sheet piles must be used to maintain trench stability.

6.1.1.1 Backhoes.
The most conventional and popular excavation technique is the backhoe. Standard backhoe excavation for shallow trenches down to 30 feet deep is the cheapest and fastest method available. The digging apparatus is staged on a crawler-mounted vehicle and consists of a boom, a dipper stick with a mounted bucket, and either cables or hydraulic cylinders to control motion (Figure 6-3). Bucket widths generally range in sizes up to 5.6 feet. Because the vertical reach of a backhoe is governed by the length of the dipper stick, backhoes can be modified with extended dipper sticks and are capable of reaching depths up to 80 feet (Day, 1996). Even greater depths are possible if benches can be excavated in which the backhoe can be located, enabling the whole backhoe to sit below grade. This can, however, be time-consuming and require a large area to be excavated to reach the required depth.

6.1.1.2 Clamshells.
Down to around 200 feet deep, a clamshell bucket can be used. A cable-suspended mechanical clamshell is a crane-operated grabbing tool that depends on gravity for accurate excavation and closure of the grab (Figure 6-4). Therefore, a heavier tool is beneficial.

Hydraulic clamshells can be equipped with a kelly bar to help guide and control the vertical line in addition to providing weight. The verticality of the excavation is controlled by the repeated cyclic lifting and lowering of the bucket under gravity. Mechanical clamshells are preferred over their hydraulic counterparts because they are more flexible in soils with boulders, can reach greater depths, and involve fewer maintenance costs. Clamshell excavation is popular because it is efficient for bulk excavations of almost any type of material except highly consolidated sediment and solid rock. It can also be controlled and operated in small and very confined areas as long as the boom can reach over the trench. Clamshell excavation, however, has a relatively low production rate compared to a backhoe. Also, worker safety can become an issue during clamshell excavation. At previous permeable barrier installations, construction sometimes involved sending a person into the trench to clear soil out of regions that were not accessible to the clamshell.

FIGURE 6-2. Temporary sheet piling separates pea gravel and iron zones in a funnel-and-gate–type permeable barrier. Permanent sheet piling was used for the funnel walls. (Groundwater flows from left to right in the picture.)

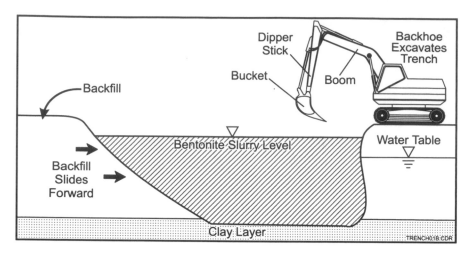

FIGURE 6-3. Conventional backhoe excavation of a slurry cutoff wall

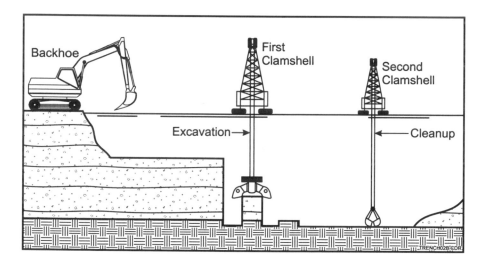

FIGURE 6-4. Trench excavation using a clamshell and backhoe

6.1.2 Caisson-Based Emplacement

A caisson is a hollow, load-bearing enclosure generally used as a retaining method for excavations (Figure 6-5). Caissons can vary in size and shape depending on the applications for which they are intended. For the purpose of emplacing

FIGURE 6-5. Emplaced caisson being augered out to make room for the iron medium

a reactive cell, a prefabricated, open, steel caisson can be used to temporarily facilitate excavation. Normally, an 8-foot-diameter (or smaller) caisson can be pushed or vibrated down into the subsurface. The smaller the diameter of the caisson, the more easily it can be driven in and maintained in a vertical position. Because caissons larger than 8 feet in diameter may not be economical for reactive cell emplacement, the flowthrough thickness and residence time in the reactive cell are limited. Therefore, at sites with wide plumes, higher levels of contamination and higher groundwater velocity, funnel-and-gate systems with multiple caisson gates typically may be used to provide adequate residence time (see Figure 6-1c).

Once the caisson has reached the intended depth, the soil within the caisson can be augered out and replaced with a reactive medium. Upon completion of the reactive cell, the caisson can be pulled straight out. Because this method requires no internal bracings, the caisson can be installed from the ground surface and completed without requiring personnel to enter the excavation. It can also be installed without having to dewater the excavation. Most small or large-sized reactive materials can be emplaced within the cell. Another advantage to this method is that it is relatively inexpensive.

During emplacement of a caisson, some soil compaction can occur along the walls of the caisson that could lower the permeability around the intended reactive cell. If the formation contains a significant amount of cobbles, the caisson may be deflected to an off-vertical position as it is pushed down, or it may even meet refusal. At one previous installation, highly consolidated sediments and cobbles created difficulties in driving in and pulling out the caisson. It may also be somewhat more difficult to drive a caisson to depths greater than about 50 feet. However, in the absence of such geotechnical difficulties, caissons have the potential to provide an inexpensive way to emplace a funnel-and-gate system and have been used for some recent permeable barrier installations.

6.1.3 Mandrel-Based Emplacement

In this method, a hollow steel shaft, or mandrel, is used to create a vertical void space in the ground for the purpose of emplacing reactive media. A sacrificial drive shoe is placed over the bottom end of the mandrel prior to being hammered down through the subsurface using a vibratory hammer. Once the void space is created, it can then be filled with a reactive medium in one of two ways. One method uses a tremie tube to simply pour the media loosely down the hole. After a desired depth is reached, the mandrel is extracted, leaving the drive shoe and media. Another way to complete the cell is to emplace wick drains, geomembranes, or geofabrics in conjunction with reactive media.

Some disadvantages to this technique include the limited size of the reactive cell, which is controlled by the size of the mandrel, typically 2-inch × 5-inch. Therefore, a series of mandrel-emplaced voids would constitute a reactive cell rather than a single insertion. Because the mandrel is hammered down using a vibratory hammer, it is possible that subsurface obstructions during installation can cause the mandrel to deviate from an intended vertical path. Also, compaction can occur around the individual voids as the mandrel is driven down, lowering the permeability of the soil.

Mandrel-based emplacement does have some advantages. It is inexpensive ($7 per ft^2 including labor and equipment for 45 feet of depth), and no spoils are generated, which minimizes hazardous waste exposure and disposal. Also, reactive material of up to 1-inch particle diameter can potentially be emplaced.

6.1.4 Continuous Trenching

Although not as common as backhoes or clamshells because of depth constraints, the continuous trencher is an option for barriers 35 to 40 feet deep. It is capable of simultaneously excavating a narrow, 12- to 24-inch-wide trench and immediately refilling it with either a reactive medium and/or a continuous sheet of impermeable, high-density polyethylene (HDPE) liner. The trencher operates by cutting through soil using a chain-saw type apparatus attached to the boom of a crawler-mounted vehicle (Figure 6-6). The boom is equipped with a trench box,

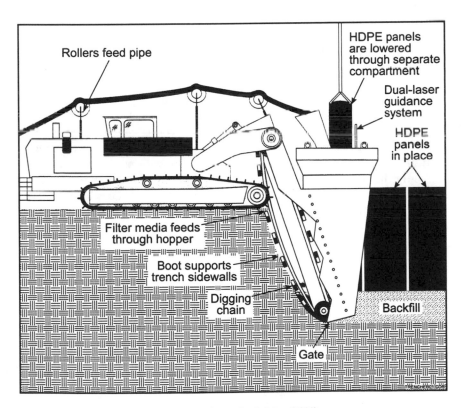

Reprinted with permission of Groundwater Control, Inc. (1996).

FIGURE 6-6. Continuous trencher in operation

which stabilizes the trench walls as a reactive medium is fed from an attached, overhead hopper into the trailing end of the excavated trench. The hopper contains two compartments, one of which can emplace up to gravel-size media. The other compartment is capable of simultaneously unrolling a continuous sheet of HDPE liner if desired.

The trencher can excavate in a water-filled trench without having to dewater or install sheet piles to temporarily stabilize the trench walls. Because the boom is positioned almost vertically during excavation, a trench slope is not created and greatly minimizes the amount of generated trench spoils. One other advantage is a fast production rate. At an installation site in Elizabeth City, North Carolina, a continuous reactive cell 150 feet long, 2 feet wide, and 26 feet deep was installed in one day (Schmithorst, 1996). Also, it is ideal for sites with constrained working space and minimizes soil disturbance to allow for work in sensitive areas. Drawbacks include a shallow depth capability and problems with excavating wet, very unconsolidated materials which may cause difficulties in bringing trench spoils to the surface. Obstructions such as large cobbles and boulders can also disrupt the sawing process. Quoted costs for this technique are between \$5 and \$12/ft^2 for emplacement, not including mobilization or reactive medium costs.

6.2 COMMERCIALLY AVAILABLE TECHNIQUES FOR FUNNEL WALL EMPLACEMENT

The design of some reactive cells may include flanking impermeable walls to aid in directing or funneling groundwater flow towards the permeable gate. The two most popular types of subsurface impermeable barriers are the steel sheet pile cutoff wall and the slurry trench cutoff wall. These subsurface cutoffs are either keyed in a confining layer to prevent downward groundwater migration, or less commonly, installed as a hanging wall to contain floating contaminants (Figure 6-7). If the presence or continuity of a confining layer is questionable, it may be possible to install a grouted impermeable bottom barrier up to 120 feet deep.

6.2.1 Steel Sheet Piles

The steel sheet piling barrier is a conventional type of subsurface barrier used in geotechnical construction applications. It is commonly used as a retention wall during excavation to prevent trench collapse and to hinder groundwater flow. It is noted for its strength and integrity and will resist hydrofracturing. The effective life range of a sheet piling wall varies between 7 and 40 years depending on the oxygen content of the soil and the corrosiveness of the contaminants (Wagner et al., 1986). Sheet piles are typically 40 feet in length but can be welded together if depths greater than 40 feet are desired. They are connected at their edge interlocks prior to being driven into the subsurface by either a drop hammer or a vibrating hammer (Figure 6-8). Sheet piles are driven in a few feet at a time along the length of the wall until they reach the desired depth. They are not feasible in very rocky

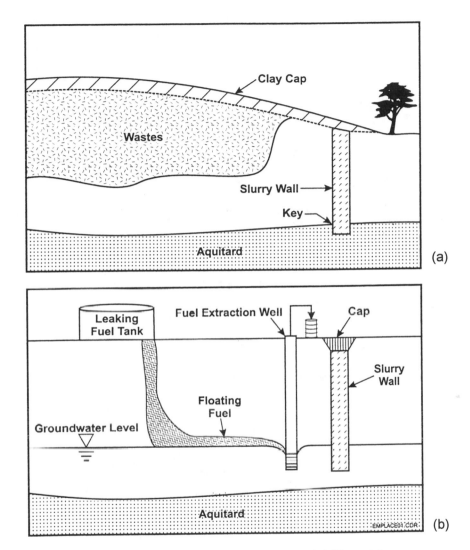

FIGURE 6-7. Types of slurry wall emplacement. (a) Keyed-in emplacement and (b) hanging wall emplacement

soils because they are likely either to be damaged during emplacement or to meet refusal. Although sheet piles have been driven down to depths of 80 feet in the past, they begin to deviate past vertical at around 60 feet. Despite sheet pile strength and integrity, conventional steel sheet pile use in environmental applications has been limited because of the leakage that occurs through the interlocks of connecting piles.

The University of Waterloo has developed sealable-joint sheet piling which has been used at several contaminated sites as cutoff walls. Very low permeabilities,

FIGURE 6-8. Sheet piles emplaced using a vibrating hammer

rapid installation, and minimal site disturbance are some features of the sealable sheet pile. This special innovation features a sealable cavity in the interlocks of connecting sheet piles (Figure 6-9). After pile sheets are driven, the joint is flushed out with jetted water prior to sealing. Also, video equipment can be lowered down

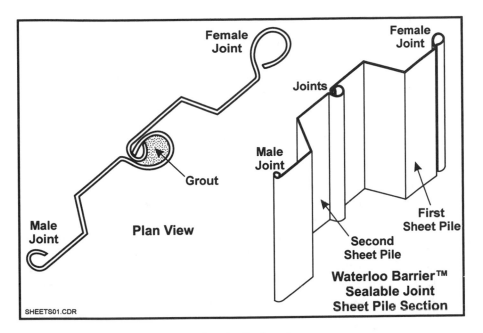

FIGURE 6-9. Sealable-joint sheet pile barrier

the cavity for visual inspection of the joint. The cavity is then sealed by grouting it from bottom to top using a tremie pipe.

Some uncertainities may occur regarding the integrity of the joint as a sheet pile is being driven. A considerable amount of friction is produced during sheet pile installation and joint flanges could weaken or be damaged, especially if greater depths are desired (Breaux, 1996). Also, the irregular shape of the individual sheet piles and the curved nature of the interlock could create some difficulties during installation. The spaces between corrugations in the sheet piles are not accessible with clamshell excavators, and this has resulted in construction personnel entering the trench to clear away these areas (Myller, 1996). The loose interlocks of connecting piles (prior to grouting) may make it difficult to drive piles in vertically without them pinching together. Experience and caution should be used to address these potential difficulties.

As with conventional steel sheet piles, the sealable-joint piles are limited to depths of 60 feet with confidence of maintaining sheet integrity and performance, but can be emplaced deeper. Rocky soils and consolidated/compacted sediments can damage pile sheets during installation and limit the types of geologic media through which the sheets can be safely driven. Use of sheet piles may be difficult in a funnel-and-gate system with caisson gates, although the difficulty of obtaining a proper seal between the funnel and reactive cell has been overcome at one existing installation through engineering modifications. The sealable-joint sheet piles currently are manufactured at only one location, in Canada, so availability could be limited.

6.2.2 Slurry Walls

Slurry walls are the most common subsurface barrier used for diverting contaminated groundwater. They are constructed by first excavating a trench under a head of liquid slurry. The slurry, which is usually a mixture of bentonite and water, helps maintain the integrity of the trench by forming a filter cake over the face of the wall. As a trench is excavated, it is quickly refilled with a mixture of cement-bentonite or a selected soil-bentonite backfill. The more common slurry walls constructed are the soil-bentonite slurry wall and the cement-bentonite slurry wall. Another, but less common, type is the plastic concrete slurry wall. All types are described in detail below.

Careful planning is critical in the design of a slurry wall. Site-specific conditions will dictate which type of slurry wall is appropriate and which is most effective. Permeability, deformability, and performance are important factors that will determine the feasibility and performance life of a slurry cutoff wall. A compatibility test can be done prior to installation to ensure that the slurry materials are suitable for the type of groundwater encountered. The trench typically is excavated by either a backhoe or a clamshell, as described in Section 6.1. Although slurry walls have been used in a variety of configurations, they are especially suited for installation as a funnel-and-gate system with caisson gates because of the ease with which the seal between the slurry wall and reactive cell can be achieved.

6.2.2.1 Soil-Bentonite Slurry Wall. Slurry walls composed of a soil-bentonite mixture are by far the most commonly used cutoff walls for environmental applications. They are the least expensive to install, have very low permeabilities, and are chemically compatible for withstanding various dissolved-phase contaminants. The construction of the wall is fairly straightforward (Figure 6-10). The bentonite slurry is introduced into the trench as soon as excavation begins. Excavated backfill can be mixed with water and bentonite. Once the trench reaches the desired depth and a sufficient length has been excavated, mixed backfill can start being pushed back into the trench. It is important to ensure that the backfill is uniformly mixed and liquid enough to flow down the trench slope. The backfill should not flow past the trench slope where it could interfere with the ongoing excavation. However, if it does not flow enough, it can start to fold over and create pockets or voids of high permeability.

Although some factors limiting the installation of a soil-bentonite slurry wall can be overcome through careful engineering, one that cannot is space availability. It is necessary to have ample work space for adequate mixing of excavated backfill and the collection of unused trench spoils.

6.2.2.2 Cement-Bentonite Slurry Wall. Some field sites may have limited work space and not allow room for mixing the excavated backfill. Another option besides the soil-bentonite slurry wall in these scenarios is a cement-bentonite slurry wall. Construction of the wall involves excavation of a trench under a head of slurry composed of water, bentonite, and cement. Instead of backfilling the trench

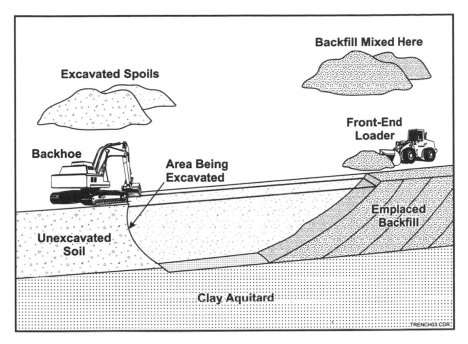

FIGURE 6-10. Cross-section of a soil-bentonite slurry trench, showing excavation and backfilling operations

with mixed soil, as in the case of a soil-bentonite wall, the slurry is left to harden and will form a wall with the consistency of a stiff clay.

The use of cement-bentonite slurry walls in environmental applications is limited for various reasons. They are more expensive to install than other slurry walls because a large amount of cement is needed to fill the trench. Also, because the excavated soil is not used as backfill, it will need to be disposed of at additional cost. Moreover, because the cement-bentonite slurry wall does not contain many solids, the wall is composed mostly of water and therefore has a higher permeability and is more prone to permeation by contaminants. Advantages of the cement-bentonite slurry wall include greater strength and the ability to be installed in areas with extreme topography.

6.2.2.3 Plastic Concrete Slurry Wall. The plastic concrete slurry wall is a variation of both the soil-bentonite and cement-bentonite slurry walls. It is composed of a mixture of water, bentonite, cement, and aggregate that hardens to form a wall with significantly greater sheer strength, yet remains flexible. The plastic concrete slurry wall is constructed in paneled sections that are individually excavated under a bentonite slurry. Once a panel is excavated, the plastic concrete is poured with a tremie pipe into the panel to replace the bentonite slurry and is left to harden. The plastic concrete slurry wall is used in applications where strength and deformability are desired. It has a relatively low. permeability, and based on limited data, may be more resistant to permeation by contaminants.

6.2.2.4 Composite Barrier Slurry Wall. This multiple-layer barrier offers three walls of defense, each with increasing chemical resistance and lower permeability. It is composed of an outer 1/8-inch-thick bentonite filter cake, a 1- to 2-foot-thick soil-bentonite, cement-bentonite, or plastic-concrete middle backfill, and an inner 100-mil HDPE geomembrane (Figure 6-11). The HDPE has a permeability of 1×10^{-12} centimeters/second. Installation of the composite barrier starts with excavation of a trench under a bentonite and/or cement slurry. Because the slurry maintains trench wall stabilization, excavations greater than 100 feet are possible; however, the difficulty of emplacing the HDPE liner to those depths and the high cost of deep emplacement has resulted in restricting the use of HDPE to 50 feet (Cavalli, 1992). The geomembrane envelope is then installed vertically in sections into the slurry trench by either mounting it onto a detachable, removable frame, pulling it down using weights affixed to the membrane bottom, or "driving" it down using a pile driver. Once the HDPE is in place, the trench can be backfilled on either side of the membrane. The inside of the geomembrane then can be filled with a drainage system in which sampling points can be installed to monitor the performance of the system.

Advantages of the composite barrier include a very low permeability, high resistance to degradation, option to install a monitoring system within the membrane, and ability to isolate and repair sections of the wall without removing the entire membrane envelope. Excavation of the trench is limited by the types of geologic media the particular excavator can tolerate. Backhoes, clamshells, and trenchers are successful in excavating most unconsolidated soils, but clamshells can also remove boulders, if necessary.

6.3 INNOVATIVE EMPLACEMENT TECHNIQUES

In addition to the emplacement techniques that have been used at permeable barrier sites in the past, several techniques have been used in other geotechnical applications and may merit serious consideration for permeable barriers. Because excavation equipment is not involved, these innovative techniques have considerable potential to reach greater depths and minimize health and safety hazards. However, because these techniques involve specialized equipment, they can be more expensive to operate and maintain than conventional means. Types of innovative emplacement techniques discussed include jetting, emplaced hydraulic fracturing, and deep soil mixing.

6.3.1 Jetting

Impermeable barriers for funnel walls can be emplaced using specialized equipment that is commonly used to inject a water and bentonite and/or cement

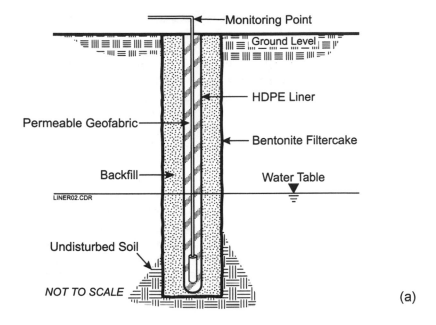

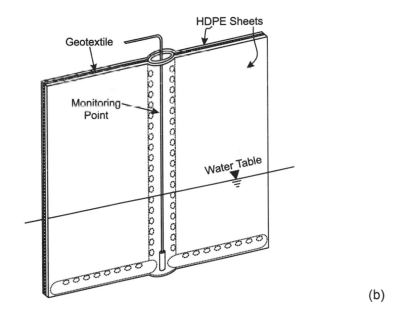

FIGURE 6-11. Composite barrier design. (a) Monitoring well cross section, and (b) section of HDPE liner envelope

slurry directly into the soil. One such method is jetting (jet grouting), in which "soilcrete" (grouted soil) columns in series form an impermeable barrier. This technique, however, involves injecting grout at high pressure through the nozzle(s) of a drill stem as it is raised up through the soil. The high-pressurized grout displaces most of the soil and can form a barrier up to depths of 130 feet. Depending on whether or not the drill stem rotates as it is raised, the resulting barrier can be either a grouted column or a thin diaphragm wall (Figure 6-12).

A systematic approach is used when injecting an impermeable funnel wall of soilcrete columns. Usually two or three rows of overlapping, interlocking columns can form an effective barrier. If three rows are desired, the two outer rows are injected first with columns emplaced in an alternating fashion. After the two outer rows are completed, the middle row of columns is injected in a similar fashion, ensuring complete contact with the columns of the outer rows (Figure 6-12a and b).

Three variations of this method are possible depending on how many of the three jetting nozzles on the drill rods are used. A single-rod system will inject only cement-bentonite grout through one nozzle into the soil. The double-rod system uses two nozzles to inject both cement-bentonite grout and compressed air. Cement-bentonite grout, compressed air, and water are injected through three ports in the triple-rod system. The added injected pressurized air and water in the latter two methods act to cut through the soil and displace it to the surface. For the purpose of creating columns, all three methods will form soilcrete columns of different sizes and will displace various amounts of spoil to the surface, depending on which method is employed and the cohesiveness of the soil. Thin diaphragm walls are formed by jetting from two nozzles, creating two halves of the wall on either side of the borehole (Figure 6-12c). The spoils that are generated typically are lighter, fine-grained clays and silts. Coarser sands and gravels remain to mix with the injected grout. Typically, a soilcrete column or diaphragm wall is about 50 percent soil and 50 percent grout in composition. If a soil is predominately composed of fines, most of the material will be displaced to the surface and a grout-rich barrier will be formed. Conversely, a predominately coarse-grained soil will displace a smaller amount of spoils and inject less grout into the soil matrix.

A single-rod system will expel a minimal amount of spoils back up the drill-hole and form columns 1.3 to 4.0 feet in diameter, depending on the soil type. The cutting action of water and compressed air in the triple-rod system can form larger columns 1.6 to 10 feet in diameter and probably will expel more spoils to the surface. A thin diaphragm wall can range in thickness from 4 to 6 inches near the nozzle to 12 to 18 inches at the furthest extent, because the spray pattern usually is fan-shaped. The lateral extent of the wall can be from 6 to 13 feet, depending on whether multiphases or multifluids are used.

Instead of injecting pressurized grout, jetting may possibly be used to inject a reactive medium, such as granular iron, for the purpose of emplacing a reactive cell. Because the injection process expels soil fines to the surface, replacement with reactive media would increase the permeability of the column. Unless the jets are modified to inject larger particles, the reactive media, such as iron, would have to be clay-sized and be suspended in a revert (biodegradable slurry). Micropowder

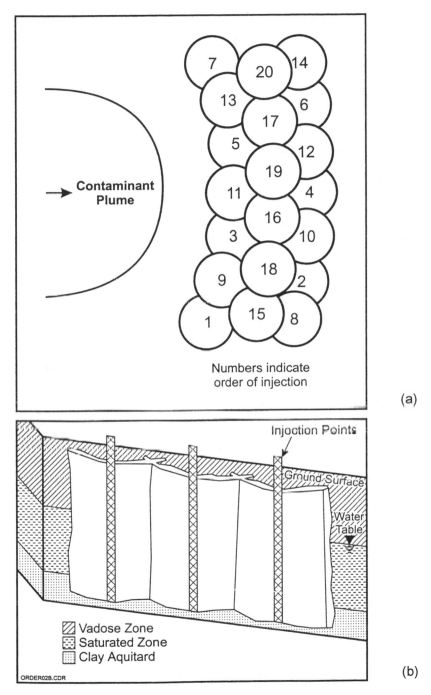

FIGURE 6-12. Diagram of (a) geometric layout of grouted injection holes, and (b) vertical thin diaphragm walls

reactants are available and would be ideal to use, but little is known about them (Gorsky, 1996). Typical problems could be increased wear on the machinery and blocking up of pipes and hoses depending on the size of the injected media. However, injection using the triple-rod system could alleviate machinery wear because this method uses lower pumping rates. More reactive media could also be injected using the triple-rod system. Injection pumps have to be properly selected to avoid wear due to abrasive media.

6.3.2 Emplaced Hydraulic Fracturing

Hydraulic fracturing is the intentional fracturing of a subsurface formation using pumped water and air under high pressures. It has been widely used in the petroleum industry as a means to increase either the delivery or recovery of petroleum hydrocarbons from low-permeability reservoirs by creating subsurface conduits for fluid flow. The technique has recently been adopted for environmental applications as a way to emplace in situ permeable barriers. A series of horizontally stacked fractures 12 to 15 meters in diameter can form an effective reactive zone to intercept and treat downward migrating contaminants.

Hydraulic fracturing begins by slowly pumping a fracturing solution (usually composed of water and guar gum) into a sealed portion at the bottom of a cased borehole. As confining pressures are exceeded in the borehole, fractures will open and propagate out laterally from an initiation point previously notched out of the casing. A fracture-fill slurry composed of a reactive medium, such as iron powder and guar gum, can then be injected into the fracture to form a reactive treatment zone. Hydraulic fracturing generally creates fractures that are only up to 3 inches thick, so more than one fracture may be required to attain the desired residence times within the treatment zone. Some advantages to this technique include the ability to emplace a barrier to a depth greater than 80 feet. Also, fracturing causes minimal site disturbance, does not generate contaminated soils, and is inexpensive. Some drawbacks of emplacement by hydraulic fracturing include difficulty in controlling the fracture direction and the limited soil conditions in which it can be used effectively, predominantly in overconsolidated sediments.

6.3.3 Deep Soil Mixing

Deep soil mixing uses special augers in series, equipped with mixing paddles that mix up soil as they rotate. Simultaneously, a bentonite slurry is injected through a hollow drill stem as the augers retreat back to the surface (Figure 6-13). An impermeable wall is formed by successive overlapping penetrations made with the deep soil mixer, resulting in a series of hardened, soilcrete columns. Typically, 40 to 60 percent of each soilcrete column is composed of grout.

Depths of up to 120 feet can be obtained using this method, and permeabilities approaching 1×10^7 centimeters/second are attainable. This method is generally employed in situations where generation of spoils on the surface is not desirable. It is best used in soft soils, yet special attention should be given so that injection

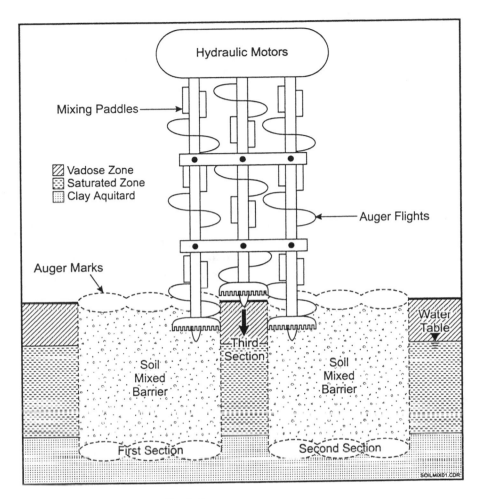

FIGURE 6-13. Deep soil mixing

does does not cause hydrofracturing of the soil, which can easily be done in soft soils. Generally, deep soil mixing is less expensive than jet grouting and has a higher production rate.

Although it has never been done commercially, it may be possible to use deep soil mixing to inject a reactive medium for the purpose of creating a reactive cell. However, because deep soil mixing does not completely replace soil with the reactive medium but rather mixes them together, only about 40 to 60 percent of the reactive medium would be present in a completed column. Increased permeability occurs as the soil mixing process fluffs up the soil matrix, yet with time, compaction due to overburdening will reduce it (Burke, 1996). The injected reactant could be equivalent to fine sand-sized particles, but would have to be suspended in a revert (biodegradable slurry) to be injected. Because the slurry is injected using piston-driven cylinder pumps, several factors should be considered when deciding on the reactant particle size. The abrasiveness of the reactant can cause considerable wear and tear on the pumps, which can increase operation and maintenance costs significantly. Also, the reactant needs to be in suspension if it is to be injected in an efficient manner. One additional factor requiring consideration is the possibility of residual slurry coatings affecting the reactivity of the reactive medium (ETI, 1996).

6.4 CONSTRUCTION QUALITY CONTROL (CQC)

The effectiveness and long-term performance of either a permeable or impermeable barrier depends on the level of construction quality control that is implemented. Appendix D addresses the CQC issues involved with the barrier technologies more commonly used in environmental applications, such as slurry cutoff walls, deep soil mixing, and jet grouting. In addition, CQC issues for sealable-joint sheet piles are mentioned because this technique is starting to be more widely used.

6.5 HEALTH AND SAFETY ISSUES

The success of any construction application can be attributed to prior knowledge of any foreseeable hazards and the taking of careful steps to avoid them through the implementation of safety practices under the guidelines outlined by the Occupational Safety and Health Administration (OSHA). A formal health and safety plan structured to address potential site-specific hazards will be required prior to commencement of construction activities. Listed below are a few health and safety issues that must be considered during permeable barrier installation:

- Confined space entry (if anticipated)

- Knowledge of location of existing utilities, including overhead or buried power lines, sewer lines, phone lines, and water pipes

- Types and concentrations of contaminants involved, which will dictate the type and level of personal protective equipment (PPE) required

- Use of heavy excavating equipment, requiring use of a hard hat, steel-toed boots, safety glasses, gloves, and hearing protection

- Trench entry, which may be necessary for visual inspection of important CQC issues, such as if excavation is keyed into a confining layer correctly or if buried utilities hinder use of mechanical excavation equipment

- Trench entry could be required to clear out the spaces inside the corrugations of sheet piles that are not reachable by clamshell excavators.

6.6 WASTE MINIMIZATION

Exposure to contaminated trench spoils is likely to occur during the emplacement of a subsurface barrier. The generation of hazardous or nonhazardous waste can be minimized through careful selection of an emplacement technique that involves either no generation of contaminated spoils or generation of only minimal amounts. Sometimes design factors will dictate that a barrier be constructed in uncontaminated soil downgradient from a contaminant plume, eliminating the problem of dealing with hazardous waste. The opposite scenario could occur, requiring excavation of soils within a contaminant plume. In any event, the amount of trenching and disposal of spoils should be planned before selecting an emplacement technique.

CHAPTER 7

MONITORING THE PERFORMANCE OF A PERMEABLE BARRIER

Once the permeable barrier has been designed and installed, the system will have to be monitored as long as the plume exists. Monitoring is done to achieve the following objectives:

- Ensure that the plume is being adequately captured and treated. Also, it is important to analyze the downgradient aquifer water to determine if the barrier has had any adverse effects on groundwater quality. The type and frequency of monitoring required to achieve this objective are usually decided during discussions between the site manager and the regulators. The monitoring for this objective falls into the category of compliance monitoring required to ensure human health risk reduction and environmental protection.

- Determine how well the installed barrier meets design specifications. During the early stages of operations, the site manager will be able to determine causes for the potential success or failure of the barrier.

- Estimate the longevity of the barrier. Over the long term, the site manager will be able to anticipate potential maintenance activities and costs.

7.1 ADEQUACY OF PLUME CAPTURE AND TREATMENT

After installation of the permeable barrier is complete, the site manager and regulators will need to know if the plume is being adequately captured and treated. From a compliance perspective, the monitoring is done to ensure that down-gradient contaminant concentrations are below target cleanup levels. This involves looking for three things:

- Potential breakthrough of contaminants or environmentally deleterious byproducts through the reactive cell

- Potential contaminant bypass around or beneath the barrier

- Potentially deleterious effects on groundwater quality due to the barrier itself.

7.1.1 Monitoring for Potential Breakthrough or Bypass of Contaminants

The type and frequency of monitoring required to achieve this critical objective are likely to be very site specific, although the ITRC Permeable Barriers Subgroup is trying to formulate guidelines for monitoring such installations. Figure 7-1 shows examples of monitoring well configurations that could be used, depending on site conditions, to monitor for breakthrough and bypass of contaminants. In Figure 7-1a, monitoring is done along the downgradient edge of the reactive cell using a row of long-screened wells. If the contaminant distribution in the plume is particularly heterogeneous with respect to depth, well clusters may be used instead of long-screen wells. Well screens are placed so as to exclude 1 foot at the top and bottom. Wells are placed a few inches inside the reactive medium rather than in the pea gravel (as in Figure 7-1b) because the pea gravel may contain some stagnant water entering from the downgradient formation. This is a concern especially if the barrier is built within the plume. Placing the monitoring wells within the reactive medium also provides a level of safety. If contaminant breakthrough is observed in these wells, there is still some reactive medium that can treat the contaminants further before the groundwater exits the reactive cell. Additional monitoring wells are placed at the two ends of the barrier to monitor for contaminant bypass that could result from inadequate flow capture.

Figure 7-1c shows another possible arrangement of monitoring wells. In this arrangement, monitoring wells are placed in the downgradient aquifer instead of in the reactive cell or pea gravel. If there is potential for flow bypass beneath or around the barrier, this arrangement could provide more information. Flow bypass beneath the barrier would occur if the barrier is not properly keyed into the aquitard or if the aquitard itself has fractures. Flow bypass around the barrier could take place if the actual hydraulic capture zone becomes smaller than designed or if the plume shape changes over time. Downgradient water quality could be monitored over increasing distances from the barrier to evaluate flow remixing and geochemical parameter rebound.

The required frequency of compliance monitoring will be determined during discussions with the regulators. Quarterly monitoring is usually required at contaminated sites. In general, the monitoring frequency for permeable barrier installations is not expected to be very high. As seen in Chapter 1, unless the dissolved oxygen content of the groundwater and the flow velocity through the reactive cell are both very high, the reactive medium is consumed slowly, over a time scale of years. One safety measure would be to install another row of monitoring wells within the reactive cell at about 1 foot from the downgradient edge (see Figure 7-1d). Because the design of the reactive cell usually involves safety factors that provide for additional reactive cell thickness, this row of wells would serve as an early indicator of impending breakthrough. Figure 7-2 shows the results of monitoring conducted in several wells along the center line through the reactive cell in the direction of groundwater flow at a pilot barrier in California (Battelle, 1997). Parts (c) and (d) in this figure indicate different depths in the monitoring well cluster at each location.

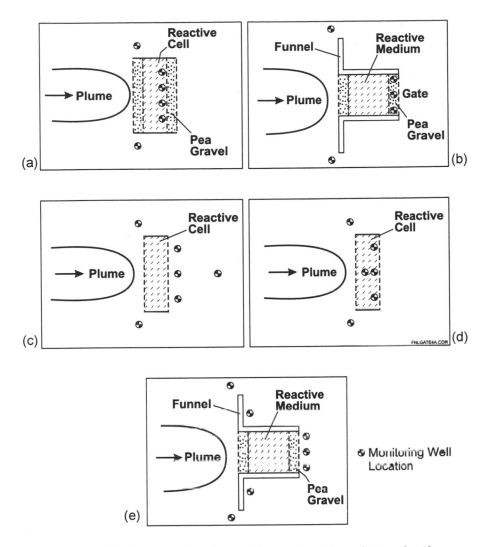

FIGURE 7-1. Various monitoring well configurations for evaluating performance of the barrier

It has been assumed that no contaminant transport occurs through the funnel walls. If there is some uncertainty regarding the impermeability of the funnel, either because of geotechnical difficulties during installation or because innovative emplacement methods were used, additional wells could be installed immediately downgradient from the funnel (see Figure 7-1e) to monitor for breakthrough.

Because monitoring costs will probably constitute the major operating cost of the barrier over the next several years, site managers will wish to optimize the number of monitoring wells and the information gained. Adequate site characterization

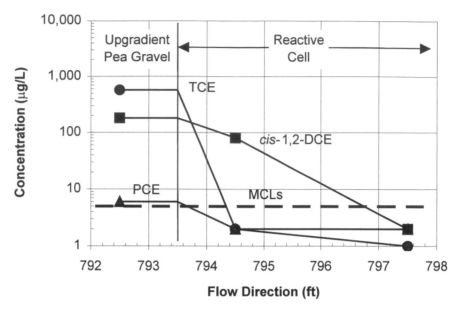

FIGURE 7-2. Concentrations of chlorinated compounds along the center line in the flowpath of existing permeable barrier (Battelle, 1997)

in the vicinity of the proposed permeable barrier location, as well as hydrologic and geochemical modeling, can assist both site managers and regulators to determine the appropriate number of monitoring wells at a given site and their locations.

Monitoring wells within the reactive cell can be either long-screen wells (for homogeneous flow) or cluster wells (for heterogeneous flow). It should be noted that, although there is homogeneous flow in the aquifer, there may be heterogeneous flow in the reactive cell depending on how well the granular reactive medium has consolidated. The monitoring wells may be constructed using 1- or 2-inch-diameter polyvinyl chloride (PVC) casing. The diameter of monitoring wells is determined based on the space available in the reactive cell and the size of the measuring instruments that will be inserted during monitoring. The wells may be installed prior to placing the granular medium in the trench, and may be supported by metal frames, which may or may not be removed as the trench is filled. Figure 7-3 shows monitoring wells being installed in a trench-type reactive cell. Figure 7-4 shows monitoring wells supported by a frame being installed in a caisson-based excavation. If an in situ groundwater velocity meter is to be used, it would have to be similarly installed during construction of the reactive cell (see Section 7.2.1).

7.1.2 Sampling and Analysis for Contaminants and Byproducts

The chemical parameters that are typically measured in the monitoring wells include concentrations of contaminants (e.g., TCE, PCE, etc.) and potential toxic

FIGURE 7-3. Installation of monitoring wells in the reactive cell and pea gravel for a trench-type permeable barrier

byproducts (e.g., *cis*-1,2-DCE, VC, etc.). Sampling and analytical techniques for monitoring wells located in the aquifer are similar to those described in Chapter 2 for site characterization. However, special precautions may be required while sampling monitoring wells located within the reactive cell or pea gravel.

When collecting groundwater samples from the reactive cell or pea gravel, traditional methods involving purging several casing or pore volumes of water prior

FIGURE 7-4. Monitoring wells supported by a frame are inserted in a caisson-type permeable barrier

to collection should be avoided. Such practices may capture water that represents a significantly lower residence time in the reactive cell. Rapid withdrawal of a water sample by any sampling method may draw water quickly from the upgradient direction, and such water may have been incompletely treated by the reactive medium. Analyzing a mixture of water from locations partially outside of the monitoring well screen could suggest higher levels of the target analytes than actually exist. Alternative sampling methods that are expected to yield more representative water samples in a permeable barrier have been discussed by Warner et al. (1996) and Kearl et al. (1994). Examples of potentially favorable groundwater sampling techniques include the following:

- Purge small volumes of casing water (micropurging) before collecting the sample. Water is extracted at very low rates for both purging and sampling to minimize disturbance of the pore water. This prevents groundwater that has not had sufficient residence time from being drawn into the sample.

- Use packers to isolate nonrepresentative casing water from the flow system.

- Sample symmetrically along the flow direction to avoid setting up artificial gradients.

- Progressively remove samples from the downgradient direction toward the upgradient direction to minimize potential cross-contamination by the sampling equipment.

Before any groundwater sampling is done it is essential to develop a Quality Assurance Project Plan (QAPP) that addresses the QA requirements of the investigation. U.S. EPA's *Pocket Guide for the Preparation of Quality Assurance Project Plans* (U.S. EPA, 1989) is a useful resource for preparing a QAPP. U.S. EPA utilizes a four-tiered project category system with Category I being the most stringent and Category IV being the least stringent. Category I plans are for projects that are of sufficient scope and substance that their results could be used directly, without additional support, for compliance or other litigation. Because such projects are intended to withstand legal challenge, Category I QA requirements are the most rigorous and detailed. Category II plans are for projects that are of sufficient scope and substance that their results could be combined with the results of other projects of similar scope to produce narratives that would be used for rulemaking, regulation making, or policy making. Other projects that do not fit these criteria, but have high visibility, would be included in this category. Category III plans are for projects that produce results that will be used for evaluating and selecting basic options, or performing feasibility studies. Most treatability studies, field pilot, and full-scale barriers would be evaluated in Category III. Category IV plans are the least stringent.

7.1.3 Monitoring Downgradient Water Quality

In addition to monitoring for breakthrough or bypass of target contaminants (and potential toxic byproducts), it is important to monitor downgradient aquifer wells for potentially undesirable characteristics, due to secondary effects of the barrier. Depending on the site geology and geochemistry, some water quality parameters that may be affected by the barrier are soluble iron (primarily ferrous), hardness (calcium, magnesium, and alkalinity), sulfate, DO, pH, and Eh. High levels of soluble iron are undesirable from a water quality perspective, because of staining that can be caused by oxidation of the ferrous iron after the water is exposed to atmospheric air. However, high soluble iron concentrations can occur only in acidic or anoxic groundwaters. In neutral to slightly alkaline water with DO above 0.01%, most available iron will oxidize to ferric species and precipitate as a very-low-solubility solid. Generally, hard water will diminish in hardness due to precipitation of calcium and magnesium carbonates within the reactive cell. However, water hardness usually is not considered to be a critical water quality parameter. If sulfur-reducing microorganisms are present, the sulfate concentration may decline due to reduction to sulfide (bisulfide) and precipitate as FeS or other low-solubility solids. In most cases, water exiting the reactive cell will have negligible DO, moderately alkaline pH (typically 8 to 11), and low Eh (< 0). However, these parameters may be quickly restored to their original values upgradient of the barrier by reaction with soil minerals and air exchange with the atmosphere. At sites with a low air permeability cover, such as pavement on the ground surface, DO concentrations may remain low, thus favoring iron solubility.

7.2 DETERMINING IF THE BARRIER MEETS DESIGN SPECIFICATIONS

A monitoring plan may be needed to determine if the barrier meets design specifications. In the short term, if the barrier is not performing according to design goals, additional monitoring can help the site manager identify the reasons. Several different monitoring activities and tests can be conducted to evaluate this objective at the discretion of the site manager:

- Estimating residence times in the reactive cell. This generally involves measurement of the groundwater velocity or hydraulic conductivity distribution in the reactive cell.

- Estimating hydraulic capture zone. This generally involves taking water-level measurements in wells upgradient of the permeable barrier.

7.2.1 Estimating Residence Time Distribution in the Reactive Cell

Degradation of halogenated hydrocarbons is controlled by rate-dependent processes taking place in the reactive cell. Therefore, residence times (the amount of time that the water is in contact with the reactive medium) affect the degree to which susceptible groundwater contaminants are degraded. Groundwater flow velocity measurements within the reactive cell provide information pertaining to residence time. In-well or in situ flowmeters may be used to monitor for spatial or temporal changes in flow velocity (Ballard, 1996). In-well groundwater velocity probes can be inserted in 2-inch or larger monitoring wells. One probe can be used to take measurements from several wells. The in situ velocity meter, on the other hand, is generally installed in the center of the reactive cell and stays there to provide a 3D groundwater velocity vector for about 1 year. The in situ meter provides an average velocity measurement in a 1-m^3 region.

Residence time is actually a range rather than a single time because of potential heterogeneity within the barrier. Heterogeneity can decrease the overall effectiveness of the reactive cell by accelerating flow at preferential locations within the cell and thus decrease contact time between the groundwater and reactive medium. Heterogeneity increases hydrodynamic dispersion, which can promote breakthrough of contaminants. Heterogeneous flow can be caused by several factors, such as differential compaction of the iron fines, development of corrosion products on reactive medium surfaces (hydrous oxides), and precipitation of secondary minerals in the interstitial pore space (e.g., calcite, siderite, brucite). Any evidence in the breakthrough of primary organic contaminants or byproducts may indicate development of heterogeneity.

Because heterogeneities can develop within the reactive cell, it is necessary to monitor for indications of chemical changes within the cell. Chemical changes can be monitored by measuring the concentrations of contaminants and native inorganic constituents (e.g., Ca, Mg, alkalinity) in monitoring wells within the reactive cell and the surrounding regions. Monitoring of the field parameters pH, Eh, and DO is very important because they can be used to determine whether conditions are conducive to the formation of inorganic precipitates. In addition, these field parameters indicate whether conditions in the cell are optimal for reductive dechlorination to occur.

Tracer tests with a conservative tracer can be used to evaluate potential heterogeneities in the reactive cell. Sodium bromide has been found to work well in an iron reactive medium when a retardation factor of 1.2 is incorporated (Sivavec, 1996; ETI, 1996). During preliminary site characterization, the levels of bromide in the native groundwater should be measured. Elevated levels of bromide in the native groundwater would make the tracer test more difficult, because a larger concentration of sodium bromide tracer would be required. At high concentrations (greater than 1 percent), bromide may be subject to a density gradient as it travels through the aquifer or reactive cell. The resulting path of the tracer, then, may not be the same as that of the organic contaminants. One advantage of a bromide tracer is its ability to be continuously monitored using downhole, ion-selective electrodes. Continuous monitoring with such probes increases the probability of capturing the tracer peak and reduces labor costs. Ion-selective probes are expensive, but their cost could be justified by reduced labor requirements and increased chances of success. Field application of tracer tests for evaluating permeable barriers has not been very successful in the past due to a variety of reasons (Focht et al., 1997). Difficulties in ensuring the success of tracer tests result from the high cost involved in obtaining adequate sampling density (number of monitoring wells), and the limitations of monitoring instruments. Other, possibly less expensive, methods of determining heterogeneities and conductivity distribution include the use of in-hole groundwater velocity probes, slug tests, and pump tests in the reactive cell. The in-hole or in situ velocity meters, especially, hold a lot of promise.

7.2.2 Estimating the Hydraulic Capture Zone Size

The size of the hydraulic capture zone is another important criterion for meeting design specifications. The permeable barrier system is designed to capture the plume and direct it through the reactive cell. This activity evaluates how well the capture is being achieved on the upgradient side of the barrier. In simulated homogeneous and isotropic aquifers, groundwater modeling studies have shown that the capture zones are symmetrical and that their size is based on the permeability contrast between the reactive cell and the aquifer medium (Starr and Cherry, 1994). However, as discussed in Section 5.1.2, the groundwater flow modeling at a heterogeneous site (Battelle, 1996) showed that subsurface channels cause the capture zones to be substantially asymmetrical, both laterally and vertically. This example illustrates the dependence of capture zones on aquifer heterogeneities and the need for detailed site characterization prior to permeable barrier placement. This can aid in designing an appropriate monitoring scheme for estimating the capture zone size through field measurements and modeling.

The predictions made through groundwater modeling can be validated by field determination of the capture zones and flow field in the immediate vicinity of the permeable cell system. Insight gained through field observations can be useful in evaluating the design configuration of the permeable cell. This information may be useful also in cases where a pilot-scale permeable cell will be upgraded to a full-scale system in the future. The contribution of the funnel walls in increasing the flow through the gate also can be evaluated. The following activities can be conducted to evaluate the hydraulic capture zone size:

- Install several monitoring points upgradient of the wall. Figure 7-5 shows an example of a monitoring well arrangement that could be used to evaluate the hydraulic capture zone width. The water-level data from these wells should be used to determine the groundwater flow field. In addition, the borehole logs from these wells can be used to enhance site characterization with data regarding aquifer heterogeneities.

- Perform slug tests in upgradient wells to determine the K distribution in the aquifer and to support any detailed capture zone models. Slug

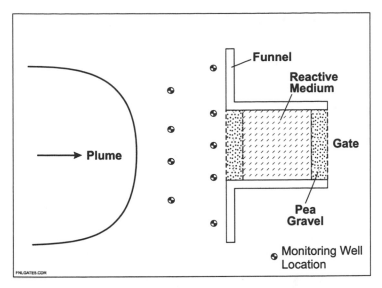

FIGURE 7-5. Example monitoring well configuration to evaluate hydraulic capture zone

tests in the reactive cell are generally ineffective because of the high porosity and conductivity of the granular medium.

- Conduct groundwater velocity vector measurements with downhole probes. These probes can measure the magnitude and direction of the groundwater flow and have the potential to indicate, at a given point, whether water is moving toward the reactive cell (or gate) or away from it. It is currently unclear how localized these measurements are, and whether they indicate flow in the bulk aquifer or a very local pore level flow. Also, at lower flowrates, such probes may not be very accurate.

- Perform tracer tests upgradient of the permeable barrier to determine flowpaths, capture zones, and flow velocities and to validate the model predictions. A relatively conservative tracer such as bromide may be used.

Tracer tests are difficult to conduct successfully without installing a large number of wells. Even wells as close as 5 feet apart can miss the tracer peak as it passes through. In addition, if the groundwater velocity at the site is very slow, it may take a long time (and higher costs) to conduct each tracer injection test. In general, estimating the hydraulic capture zone at many sites may be a difficult process.

7.3 ESTIMATING THE LONGEVITY OF THE BARRIER

The reactive cell causes dechlorination of groundwater contaminants by a reduction process in which the iron metal is consumed and iron oxide compounds are formed. As these and other compounds precipitate on the reactive metal surfaces, the reactive cell may become less effective in capturing and treating contaminated groundwater. It is of interest to determine how performance will be affected over time and to understand exactly what processes are responsible for a decline in performance. In addition, it is important to be able to predict how long the reactive cell will remain useful, both for planning purposes and to annualize cost. Some methods that may be used to assess whether barrier performance is deteriorating are as follows:

- Comparison of time series measurements of groundwater and hydrologic parameters

- Analysis of core samples within the reactive cell.

The techniques described in Section 7.1 to monitor breakthrough of contaminants and byproducts during each sampling round can be applied during multiple sampling events to determine whether performance of the reactive cell is changing over time. To determine whether temporal changes occur, groundwater samples should be collected from the same monitoring wells at regular intervals during the entire course of the evaluation. The data collected can be subjected to statistical techniques to determine whether variations are time-dependent or are random fluctuations. In addition, the ratio of concentrations in two wells in a particular flowpath may be tracked over time. Any consistent increase in the ratio of the upgradient well concentrations to the downgradient well concentrations would indicate a loss in performance. If any of the data represent time-dependent variables, it would be useful to extract rate information about the variables. Rate information could be extrapolated to predict the useful lifetime of the cell.

It is expected that the cell's peak performance will occur when the reactive metal particles are still relatively new and few mineral precipitates have deposited in the interstitial pore spaces. As the reactive medium oxidizes (corrodes) and the pore spaces become clogged by mineral precipitates, the performance of the reactive cell should decline. To better understand the processes that affect the longevity of the cell it is desirable to periodically monitor groundwater parameters that are related to changes in performance. Some key groundwater parameters that should be monitored are pH, Eh, and DO. Similarly, chemical species that may react in the cell include Ca, Fe, Mg, Mn, Al, Ba, Cl, F, SO_4^{2-}, and HCO_3^- (alkalinity); significant redox-sensitive elements include Fe, C, S, and N.

Geochemical modeling codes (see Section 5.2) can be used to evaluate the potential for precipitation in the reactive cell. Inverse modeling codes are especially suited for evaluating probable reactions in the reactive cell, given the concentrations of the various chemical species observed in upgradient and downgradient monitoring wells (before and after the reactive cell). Even in the absence of computerized modeling, just an arithmetical comparison of parameters, such as Ca and Mg, before and after the reactive cell can provide a valuable indication of potential reactive cell reactions.

If there are indications of a loss in performance due to precipitation, a few core samples taken from within the reactive cell can be evaluated. The cores could be taken in a vertical direction from the first few inches in the upgradient section, from the middle, and from the last few inches in the downgradient section of the cell to get adequate spatial information about possible changes in the reactive medium. A horizontal core through the upgradient pea gravel-reactive cell interface also could be collected to provide information on front-end corrosion of the iron due to DO in the groundwater. It will be necessary to prevent oxygen from contacting the cores prior to analysis. The boreholes can be backfilled with fresh reactive medium to restore the integrity of the cell.

Optical microscopy can be performed for visual inspection of the core samples. Microscope imaging techniques can be applied to characterize the iron oxide coatings, mineral precipitates, and any other particulate matter. Scanning electron microscopy (SEM) can be used if the benefits of backscatter electron images and elemental mapping are needed. In addition, the elemental composition can be determined by energy-dispersive x-ray spectroscopy (EDS) using an SEM, or by wave-dispersive spectroscopy (WDS) using an electron microprobe. Powder x-ray diffraction (XRD) techniques can be used to identify mineral precipitates by their crystal structures. Because iron hydroxides formed at low temperatures tend to be amorphous, they may not be detectable by XRD. If necessary, the particle number and particle-size distribution (dimensional area, and aspect ratio) can be determined using an imaging system interfaced to an optical or electron microscope. The core samples also can be evaluated for biological growth (see Appendix A, Section A.3).

The hydraulic capture volume of the permeable barrier might decrease if the hydraulic conductivity of the reactive cell is reduced due to precipitate or bacterial buildup on the reactive medium. The following activities could be conducted to evaluate changes in the hydraulic performance of the barrier over time:

- Conduct hydrogeologic modeling to develop cause-effect scenarios, including the effect of lower reactive cell conductivity on the capture zones, shift in flow divides, and change in flow volumes.

- Monitor the water levels periodically in all wells and continuously in a few wells to determine and interpret any changes in water levels over time. It is expected that any significant decrease in reactive cell conductivity will cause a water-level buildup upgradient of the cell. The water-level data also can be used to determine any changes in flow fields over time.

- If a significant change in water levels and flow patterns over time is observed, conduct slug tests within the reactive cell to determine whether the water-level changes are related to a decrease in hydraulic conductivity of the reactive cell. It is possible that the high conductivity of the reactive medium in the early stages of monitoring will result in extremely high response times during slug tests. Therefore, care should be exercised to record water-level changes at the highest possible frequency of the data loggers.

- If significant changes in water levels and conductivity are observed, the tracer test could be repeated upgradient of the cell to evaluate the effect on the capture zone and flow velocities. However, tracer tests are difficult to conduct and water-level measurements in upgradient wells may still provide a reasonably good estimate of the capture zone.

CHAPTER 8

PERMEABLE BARRIER ECONOMICS

The economical benefit of permeable barriers has been an important driving force behind the widespread interest in this technology. At chlorinated solvent-contaminated sites, where the plume could continuously be generated for several years or decades, a passive technology that requires almost no annual energy or labor input (other than for site monitoring) has obvious cost advantages over a conventional pump-and-treat system. A cost-benefit approach that includes both tangible and intangible costs and benefits should be used in evaluating the economic feasibility of a permeable barrier at a given site.

8.1 CAPITAL COST CONSIDERATIONS

The capital cost of installing a permeable barrier includes the following items:

- Cost of the reactive medium
- Cost of the emplacement
- Technology licensing costs
- Cost of disposal of spoils and restoration of ground surface.

The unit cost of the reactive medium depends on the type of medium selected. Granular iron is the cheapest of the available metallic media, and therefore has been the most commonly used. Only when special site requirements necessitate the use of a different medium may other more costly options be preferred. Although initial field applications are reported to have paid up to $650/ton for the granular iron, identification of additional sources has reduced the unit cost of iron available to approximately $350/ton.

The total cost of the reactive medium is driven not only by the unit cost of the reactive metal, but also by the amount of reactive metal required. The amount of reactive metal required depends on the following site-specific factors:

- **Type and Concentrations of the Chlorinated Contaminants.**
  Contaminants that have longer half-lives require a larger flowthrough thickness of the reactive cell, and therefore incur higher cost.

- **Regulatory Treatment Criteria.** The more stringent the treatment criteria that the barrier has to meet, the greater the residence time required; the greater the residence time, the greater the thickness of the reactive cell must be, thereby increasing the cost.

- **Groundwater Velocity**. The higher the groundwater velocity, the greater the thickness of the reactive cell required to obtain a certain residence time, and therefore the higher the cost.

- **Grondwater Flow and Contaminant Distribution.** At sites where the distribution of groundwater flow or contaminants is very heterogeneous, a reactive cell of uniform thickness and extent can lead to an inefficient use of reactive medium. Emplacement of the reactive cell in zones of higher permeability or the use of funnel-and-gate configurations and pea gravel-lined cells are some of the ways in which the use of a reactive medium may be made more efficient.

The unit costs of emplacement depend on the type of technique selected, which, in turn, depends on the depth of the installation. Table 6-1 (in Chapter 6) summarizes the emplacement techniques available, the maximum depth possible for each technique, and some representative unit costs obtained from several geotechnical contractors. Although some variability in the cost of each technique represents differences in vendors, the range of unit costs is more likely driven by depth. The total cost of emplacement is based on three main factors:

- **Plume and Aquifer Depth**. For a given emplacement technique, the upper part of the cost range generally applies to the greater depths in its range.

- **Plume Width**. The greater the width of the plume, the wider is the permeable barrier required to capture it.

- **Geotechnical Considerations**. The presence of rocks or highly consolidated sediments may make it harder to drive the emplacement equipment (e.g., sheet piles or caissons) into the ground.

Given the cost difference between the emplacement techniques for the funnel versus those for the reactive cell in Table 6-1 (Chapter 6), there may be a cost trade-off between selecting a funnel-and-gate system versus a continuous reactive barrier. Disposal of spoils generated during emplacement is another cost that may vary based on the emplacement technique selected. For example, emplacement of slurry walls generates more spoils than does emplacement of sheet piles. Disposal of spoils could be costlier if the barrier has to be located within the plume, in which case the spoils may have to be disposed of as hazardous waste. Restoration of the site surface may include returning it to grade or repaving the surface for built-up sites, and is an additional cost.

For each permeable barrier application, ETI, which holds a patent for the zero-valent iron technology, charges a licensing fee of up to 15 percent of the capital costs (materials and construction costs) for the application of the technology. A limited warranty is provided by ETI guaranteeing degradation as long as the required residence time is achieved (ETI, 1996). Other patented reactive media may have similar licensing requirements. The University of Waterloo has a patent pending for the Waterloo Barrier technology of sheet piles with grout-sealed joints, although any licensing fees are included in the price of the technology.

8.2 OPERATING AND MAINTENANCE (O&M) COST CONSIDERATIONS

O&M costs for the permeable barrier include the following items:

- **Compliance Monitoring Costs.** These costs will vary from site to site depending on regulatory requirements, number of monitoring wells, and frequency of sampling.

- **Additional Performance Monitoring Costs.** If additional monitoring is desired to achieve other engineering objectives (see Section 7.2), these costs will vary depending on the objectives of each site.

- **Periodic Maintenance Costs.** Maintenance may be required if precipitates build up to a point where either the reactivity or the hydraulic conductivity of the reactive cell is significantly affected. The reactive cell may have to be flushed with reagents, or the reactive medium may have to be replaced. Indications from existing permeable barrier sites are that such maintenance would be infrequent at most sites.

There currently is no clear basis for estimating maintenance costs, mainly because their frequency of occurrence, if at all, is not known. Rule-of-thumb criteria that have been used for economic analysis at previous sites include a maintenance cost estimate that assumes 25 percent of the reactive metal will be replaced every 10 years for sites with low precipitation potential, and every 5 years for sites with high precipitation potential. Estimating the longevity of the permeable barrier and devising methods for rejuvenating or replacing it, if required, are some areas in which additional research efforts are required.

8.3 COST-BENEFIT EVALUATION

Table 1-2 (in Chapter 1) summarizes the capital costs of permeable barriers at various sites with different barrier configurations and emplacement techniques. All these sites used granular iron as the reactive medium. In evaluating the economic feasibility of a permeable barrier application, the estimated capital costs and projected operating costs of a permeable barrier may be compared to similar estimates for a pump-and-treat system or other remedial option.

Some sites have been using a 30-year time period over which to estimate costs of a permeable barriers project. The present value (PV) of the project is determined as follows:

$$PV_{project} \quad = \quad Capital\ costs + PV_{O\&M\ costs}$$

$$PV_{O\&M\ costs} \quad = \quad Sum\ of\ annual\ O\&M\ costs\ adjusted\ for$$
$$inflation\ and\ cost\ of\ capital\ over\ 30\ years.$$

A $PV_{project}$ can be calculated separately for a pump-and-treat system and for the permeable barrier option, and then compared to evaluate the projected lifetime costs of the two approaches. Sometimes the economic benefits of a permeable barrier may not be obvious. At one existing site, the site owner was able to lease the property after installing a permeable barrier only because there were no aboveground structures. Intangible benefits should also be considered in the economic analysis.

Any economic benefits of a permeable barrier application may be included in the analysis as an offset (or reduction) to capital or O&M costs. For example, if installation of a permeable barrier results in the sale of a property that was previously unsaleable because of the need to operate aboveground pump-and-treat systems, then the sale value can be used to offset capital costs. If a permeable barrier installation results in the ability of the owner to lease or otherwise put a previously unused property to productive use, the PV of the annual cash flows from the lease or other use may be used to offset the PV of the O&M costs of the barrier. Intangible benefits, such as the risk reduction achieved by emplacing the barrier, should also be considered.

8.4 COMPUTERIZED COST MODELS

The Remedial Action Cost Engineering and Requirements (RACER) System is an environmental costing program developed by the U.S. Air Force. It can estimate costs for various phases of a remediation project:

- Site Characterization Studies (Underground Storage Tank [UST] Site Assessment, RI/FS, RCRA [RFI/CMS])

- Remedial Design

- Remedial Action (including Operations and Maintenance)

- Site Work and Utilities.

The program's framework is based on actual engineering solutions gathered from historical project information, construction management companies, government laboratories, vendors, and contractors. It is designed to factor in specific project conditions and requirements based on minimal user input to generate a cost estimate. RACER Version 3.2 has a cost database created mostly from the U.S. Army Corps of Engineers' Unit Price Book and supplemented by vendor and contractor quotes. Version 3.2 has been adapted especially for permeable barrier applications. It runs through a four-step process in creating a cost estimate (Delta Research Corp., 1996).

- **Establish an Active Project Site.** When you create an estimate, you must first establish an active project site by one of three methods: (1) add a new project site, (2) select an existing project site, or (3) copy an existing project site to a new project site.

- **Calculate Site (Direct) Cost.** A project may be comprised of a single site or it may contain several sites. For each site included in the project, select and run the technologies and/or processes (cost models) that will be used to remediate the site. The costs calculated for these models are *direct costs* only (i.e., the cost does not include contractor overhead and profit, cost for contingencies, project management, or escalation).

- **Calculate Project Costs.** Having run and calculated the direct costs for all cost models included in each site for the project, complete the estimate by calculating the project costs. Project costs include costs for contractor overhead and profit, contingencies, project management, and escalation. Because these costs apply to the entire project, they need to be calculated only once, unless the direct costs are changed.

- **Run and Print Reports.** Having completed the estimate, print the reports. Up to five different reports may be printed, with varying levels of detail. For example, you may print a report exclusively for the direct costs (Detail Cost Report), or you may print a report for the total project cost (Project Cost Report).

As an option for estimating the total project cost for installing a permeable barrier system, RACER 3.2 is very detailed and includes direct costs for remedial action professional labor, sampling and analysis, remedial design, construction phases, and operation and maintenance. The estimate is calculated based on minimal input from the user. The required input includes the following:

- Site location
- Starting/completion dates for various phases of the project
- Projected hours for involved staff
- Type of barrier
- Dimensions and media used for funnel-and-gate walls.

Many default values have already been inserted and need to be carefully scrutinized. The user must be fully aware of all the different cost models available. Careful screening of the itemized cost breakdown within each individual cost model will enable the user to edit the estimate and customize it to fit his/her needs (e.g., deleting the cost for a chemical toilet if one is already provided on site).

Some major limitations to the RACER model include the following:

- Restriction to only three design types
  1. Continuous reactive barrier
  2. Funnel-and-gate system with sheet piling (for funnel walls)
  3. Funnel-and-gate system with slurry wall (for funnel wall).

- No options for innovative emplacement techniques, such as jetted barriers, sealable-joint sheet piles, geomembranes, composite walls, caissons, etc.

- Limited methods of excavation. The model assumes all trenches up to 25 feet are excavated by a backhoe, and trenches 26 to 120 feet are excavated under a slurry. It does not account for equipment such as clamshells or continuous trenchers.

CHAPTER 9

SUMMARY OF THE DESIGN AND IMPLEMENTATION METHODOLOGY

A common environmental problem facing several government and industrial installations is the presence of chlorinated solvent-contaminated soil and groundwater. Chlorinated solvents, such as TCE and PCE, were commonly used in aircraft maintenance, dry cleaning, metal finishing, and other operations. These solvents have entered the ground through leaks, spills, or past disposal practices. The U.S. Air Force alone may have more than 600 such sites at bases across the country. The U.S. EPA (1996) estimates that there are 5,000 U.S. Department of Defense (DoD), U.S. Department of Energy (DOE), and Superfund sites contaminated with chlorinated solvents. This chapter summarizes the steps involved in the evaluation, design, and implementation of a permeable barrier, and provides a quick reference for the key issues discussed in the preceding chapters.

9.1 OVERVIEW OF THE DESIGN METHODOLOGY

Unlike conventional ex situ technologies, such as pump-and-treat systems, in situ technologies are more dependent on site-specific parameters. Therefore, this book does not purport to replace the scientific judgment of the site hydrogeologist or site engineer. Instead, this document points out the important considerations and various available options applicable to permeable barriers that should be taken into account during design, implementation, and monitoring.

Because of the multitude of competing reactions and varied kinetics that are believed to take place in the reactive cell (see Chapter 1 for a discussion of various degradation pathways), stoichiometry cannot be the basis of predicting contaminant disappearance and quantifying reaction products. Therefore, the design of a permeable barrier relies mainly on good site characterization, treatability testing, and modeling. Figure 9-1 shows the steps in designing a permeable barrier for chlorinated solvents. Chapters 2 through 8 are organized along these steps. Some steps in Figure 9-1 are interdependent and may involve trade-offs.

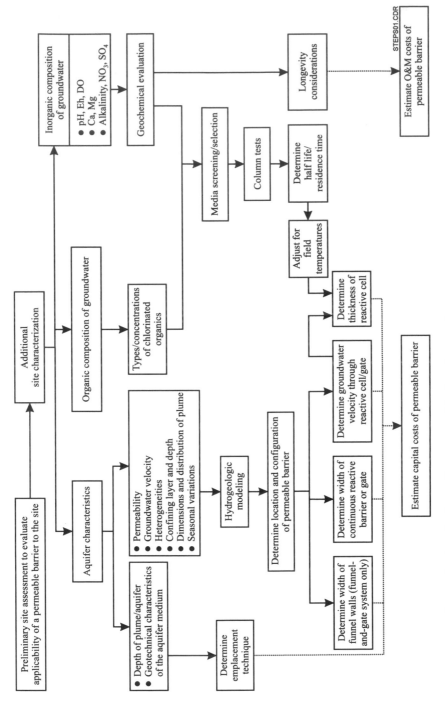

FIGURE 9-1. Steps in the design of a permeable barrier system

9.2 PRELIMINARY ASSESSMENT TO DETERMINE SUITABILITY OF A SITE FOR PERMEABLE BARRIER APPLICATIONS

Typically, the first assessment that site managers must make is whether or not the site is suitable for a permeable barrier application (see Figure 9-2). Although an unfavorable response to any of the following factors does not necessarily rule out a permeable barrier, such a response can make the application more difficult or costly:

- **Contaminant Type.** Are the contaminants of a type reported in scientific and technical literature as amenable to degradation by metallic media? Table 9-1 lists the contaminants that are currently reported as either amenable or recalcitrant to abiotic degradation with iron. An economically feasible half-life is necessary to support the application. As alternative media or enhancements are discovered, more contaminants may come within the scope of this technology.

- **Plume Size and Distribution.** Is the plume very wide or very deep? Very wide or very deep plumes will increase the cost of the application. However at least two sites currently have installed permeable barriers that are over 1,000 feet wide (see Chapter 1). In comparison with the costs of using a pump-and treat system to stop a wide plume from moving out of the site boundary, a permeable barrier may still make economic sense. Depth of the plume or depth of the aquitard may be a more significant cost consideration.

- **Depth of Aquitard.** Is the aquitard very deep? If the aquitard is very deep and the barrier has to be keyed into it, the emplacement costs could be high. For most chlorinated solvent applications, the permeable barrier will probably have to be keyed in because of the potential for underflow of contaminants. Currently, barriers can be emplaced to depths of 25 to 30 feet with standard, relatively inexpensive excavation equipment, such as a backhoe. At greater depths, more expensive methods, such as clamshell excavation, may have to be deployed. However, innovative emplacement techniques, such as continuous trenchers or the use of caisson gates, are being increasingly used at several sites to overcome depth and cost constraints (see Chapter 6).

- **Geotechnical Considerations.** Are there any geologic features at the site that may make installation more difficult? For example, the presence of consolidated sediments or large gravel or rocks may make some types of emplacement more difficult. Caissons, for instance, may bend or get caught in such formations. Aboveground

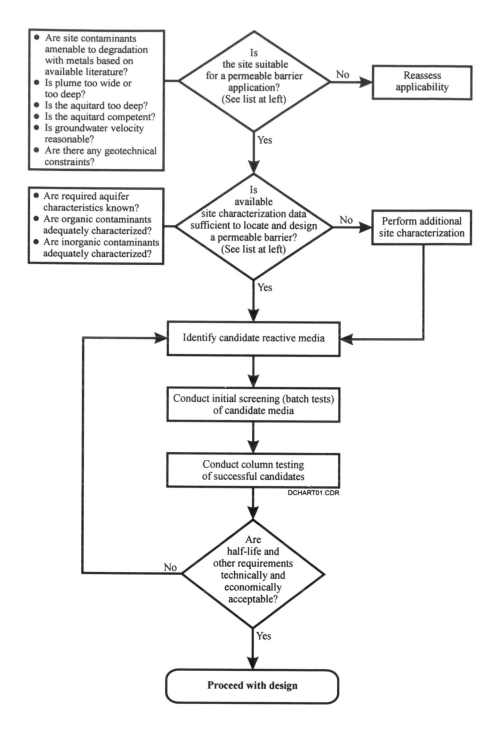

FIGURE 9-2. Decision chart for permeable barrier design activities

TABLE 9-1. Degradation rates reported as half-lives normalized to 1 m² iron surface per mL solution[a]

Compounds	Pure Iron $t_{1/2}$ (hr)	Commercial Iron $t_{1/2}$ (hr)
Ethenes		
Tetrachloroethene	$0.28^{[b]}$, $5.2^{[i]}$	$2.1\text{-}10.8^{[c]}$, $3.2^{[f]}$
Trichloroethene	$0.67^{[b]}$, $7.3\text{-}9.7^{[h]}$, $0.68^{[k]}$	$1.1\text{-}4.6^{[c]}$, $2.4^{[f]}$, $2.8^{[g]}$
1,1-Dichloroethene	$5.5^{[b]}$, $2.8^{[i]}$	$37.4^{[f]}$, $15.2^{[g]}$
trans-1,2-Dichloroethene	$6.4^{[b]}$	$4.9^{[c]}$, $6.9^{[f]}$, $7.6^{[g]}$
cis-1,2-Dichloroethene	$19.7^{[b]}$	$10.8\text{-}33.9^{[c]}$, $47.6^{[f]}$
Vinyl Chloride	$12.6^{[b]}$	$10.8\text{-}12.3^{[c]}$, $4.7^{[f]}$
Ethanes		
Hexachloroethane	$0.013^{[b]}$	—
1,1,2,2-Tetrachloroethane	$0.053^{[b]}$	—
1,1,1,2-Tetrachloroethane	$0.049^{[b]}$	—
1,1,1-Trichloroethane	$0.065^{[b]}$, $1.4^{[i]}$	$1.7\text{-}4.1^{[c]}$
1,1-Dichloroethane	Not reported	Not reported
Methanes		
Carbon Tetrachloride	$0.02^{[b]}$, $0.003^{[h]}$, $0.023^{[i]}$	$0.31\text{-}0.85^{[c]}$
Chloroform	$1.49^{[b]}$, $0.73^{[h]}$	$4.8^{[c]}$
Bromoform	$0.041^{[b]}$	
Other Compounds		
1,1,2-Trichlorotrifluoroethane (Freon 113)	$1.02^{[c]}$	—
1,2,3-Trichloropropane	—	$24.0^{[d]}$
1,2-Dichloropropane	—	$4.5^{[d]}$
1,3-Dichloropropane	—	$2.2^{[d]}$
1,2-Dibromo-3-chloropropane	—	$0.72^{[c]}$
1,2-Dibromoethane	—	$1.5\text{-}6.5^{[c]}$
n-Nitrosodimethylamine (NDMA)	$1.83^{[e]}$	—
Nitrobenzene	$0.008^{[e]}$	—
No Noticeable Degradation		
Dichloromethane,[b],[h],[i] 1,4-Dichlorobenzene[i], 1,2-Dichloroethane[c], Chloromethane[c]		

(a) Source: Reprinted with permission from Robert Gillham and John Vogan of Envirometal Technologies, Inc., Guelph, Ontario (Gillham, 1996).

(b) Gillham and O'Hannesin (1994).	(g) Mackenzie et al. (1995).
(c) Unpublished Waterloo data.	(h) Matheson and Tratnyek (1994).
(d) Focht (1994).	(i) Schreier and Reinhard (1994).
(e) Agrawal and Tratnyek (1994).	(j) Lipczynska-Kochany et al. (1994).
(f) Sivavec and Horney (1995).	(k) Orth and Gillham (1995).

structures, such as buildings, that are in the vicinity of the installation may impede the maneuverability of construction equipment.

- **Competent Aquitard.** Is the aquitard very thin or discontinuous? If so, there could be significant upward gradients across the aquitard; or if there are deeper aquifers that could be affected by a breach of the aquitard during installation of the barrier, the application should be reassessed.

- **Groundwater Velocity.** Is the groundwater velocity too high? If the velocity is high, the reactive cell thickness required to get the design residence time may also be high and the barrier could become costly. However, permeable barriers have been installed at sites with groundwater velocities as high as 4 to 5 feet/day.

9.3 SITE CHARACTERIZATION TO SUPPORT PERMEABLE BARRIER DESIGN

If a preliminary assessment shows that the site is suitable, the next issue is whether or not the available site characterization data are sufficient to locate and design the barrier. If the site information is inadequate for the purpose, additional site characterization may be required. Chapter 2, Site Characterization Data, describes the site information that is required and discusses the tools available to collect this information. The important site information required includes the following:

- **Aquifer Characteristics.** The aquifer characteristics that should be known include groundwater depth, depth to aquitard, aquitard thickness and continuity, groundwater velocity, lateral and vertical gradients, site stratigraphy/heterogeneities, hydraulic conductivities of the different layers, porosity, and dimensions and distribution of the plume. This information is required to assist in hydrogeologic modeling that will be done to locate and design the barrier.

- **Organic Composition of the Groundwater.** The types of chlorinated solvent compounds and the concentrations should be known. This information will be used to select appropriate reactive media, conduct treatability tests, and design the thickness of the wall.

- **Inorganic Composition of the Groundwater.** This information is required to evaluate the long-term performance of the permeable barrier and select appropriate reactive media. Knowledge of the presence and concentrations of calcium, magnesium, iron, alkalinity (bicarbonate), chloride, nitrate, and sulfate can be used to evaluate the potential for precipitate formation that may affect the reactivity and hydraulic performance of the permeable barrier. Field parameters such as pH, redox potential, and dissolved oxygen are also good indicators of conditions conducive to formation of precipitates.

9.4 REACTIVE MEDIA SELECTION

Once the required site characterization data have been obtained, the next step is to identify and screen candidate reactive media. Chapter 3, Reactive Media

Selection, discusses the various media available and the factors affecting their selection. The main considerations in identifying initial candidates are as follows:

- **Reactivity.** The candidate medium should be able to degrade the target contaminants within an acceptable residence time. Generally, the shorter the half-life of the contaminant with a given media, or higher the reaction rate constant, the better the media. Table 9-1 shows the ranges of half-lives of several contaminants that are degraded by iron. Any alternative medium selected should have comparable or better reactivity, unless other concerns (below) dictate a trade-off.

- **Hydraulic Performance.** Selection of the particle size of the reactive medium should take into account the trade-off between reactivity and hydraulic conductivity. Generally, higher reactivity requires lower particle size (higher total surface area), whereas higher hydraulic conductivity requires larger particle size.

- **Stability.** The candidate medium should be able to retain its reactivity and hydraulic conductivity over time. This consideration is governed by the potential for precipitate formation and depends on how well the candidate medium is able to address the inorganic components of the site groundwater. One important characteristic of the groundwater that limits precipitate formation is alkalinity, which acts as a buffer. If natural buffers are absent from the groundwater, a reactive medium that provides the required buffering capacity could be incorporated.

- **Environmentally Compatible Byproducts.** The byproducts generated during degradation should not have deleterious effects of their own on the environment. For example, during degradation of TCE by iron, small amounts of potentially toxic byproducts, such as vinyl chloride, may be generated (see Chapter 1). However, given sufficient residence time for groundwater flow through the reactive cell, these byproducts are themselves degraded to nontoxic compounds. Any alternative reactive medium selected should demonstrate similar environmental compatibility.

- **Availability and Price.** The candidate medium should be easily available in large quantities at a reasonable price, although special site considerations may sometimes justify a higher price. Currently, granular iron appears to be the cheapest and most easily available reactive medium with a price in the $350 to $400 per ton range.

At the current time, granular iron appears to be the one reactive medium that favorably meets most of the above criteria. However, alternative media that may improve reactivity (e.g., bimetallic media) and stability (e.g., iron-pyrite mixtures) are being investigated.

9.5 TREATABILITY TESTING TO GENERATE CONTAMINANT- AND SITE-SPECIFIC DESIGN DATA

Following identification of candidate reactive media, batch tests could be performed to quickly screen several candidates. If only one or two candidates have been identified, screening by batch testing could be foregone in favor of column tests. Column tests are more representative than batch tests of dynamic field conditions and provide more accurate design information. Column tests are conducted to select the final reactive medium and determine half-lives and residence times. It is recommended that column tests be performed with groundwater obtained from the site in order to generate representative design data. Half-lives calculated through column tests may require adjustments for field groundwater temperatures and the potentially lower field bulk density of the reactive medium. The flow-through thickness of the reactive cell is determined by residence time requirements and the local groundwater velocity through the reactive cell. Chapter 4, Treatability Testing, describes the methods for performing batch and column tests and interpreting the resulting data. Chapter 5, Modeling to Support the Permeable Barrier Design, describes how local groundwater velocity in the reactive cell is determined through hydrogeologic modeling.

One decision that a site manager may have to make, following column testing and determination of half-lives, is whether to do a pilot test or proceed directly to a full-scale installation. The following factors could make a field pilot installation desirable, although not necessary:

- **Complex Site**. If the site is heterogeneous and behavior of the hydraulic flow system is not well-understood, it may make sense to do a pilot test to reduce the risk of locating or installing a barrier improperly with respect to the flow system. The pilot barrier should also be installed at the location at which the full-scale barrier will most likely be emplaced.

- **New Emplacement Technique.** A pilot installation may be useful if a new emplacement technique is being used for the reactive cell or funnel walls.

- **Cost**. If the expected cost of a full-scale barrier is very high, and a pilot barrier can be installed for a small fraction of the cost of a full-scale barrier, it may be worthwhile to conduct a pilot test to reduce the risk of the larger investment. However, if the field pilot barrier is comparable in cost to a full-scale barrier, a pilot test may not be worthwhile.

9.6 MODELING TO SUPPORT BARRIER DESIGN AND DEVELOP A MONITORING PLAN

While treatability tests are being conducted, hydrogeologic modeling and geochemical evaluation of the site can begin. Hydrogeologic modeling (see Section 5.1) can be used to define many aspects of the design. Several hydrogeologic models are available for modeling the permeable barrier flow and transport system. Appendix B, Section B.1 describes the various flow and particle transport models available and their main features. Widely available and validated models such as MODFLOW and its enhancements are generally sufficient to achieve permeable barrier design objectives. Hydrogeologic modeling, along with site characterization data, is used for the following purposes:

- **Location of Barrier.** Determine a suitable location for the permeable barrier with respect to the plume distribution, site hydrogeology, and site-specific features, such as property boundaries, underground utilities, etc.

- **Barrier Configuration.** Determine a suitable permeable barrier configuration (e.g., continuous reactive barrier or funnel-and-gate system).

- **Barrier Dimensions.** Determine the width of the reactive cell and, for a funnel-and-gate configuration, the width of the funnel.

- **Hydraulic Capture Zone.** Estimate the hydraulic capture zone of the permeable barrier.

- **Design Trade-Offs.** Identify a balance between hydraulic capture zone and flowthrough thickness of the reactive cell (gate), which are interdependent parameters.

- **Media Selection.** Help in media selection and long-term performance evaluation by specifying required particle size (and hydraulic conductivity) of the reactive medium with respect to the hydraulic conductivity of the aquifer.

- **Longevity Scenarios.** Evaluate scenarios for future potential flow bypass due to reduced porosity resulting from precipitate formation. This gives an indication of the safety factors needed in the design.

- **Monitoring Plan.** Assist in planning appropriate monitoring well locations and monitoring frequencies.

Geochemical evaluation (see Section 5.2) of the site can also commence while treatability tests are in progress, although knowledge of the inorganic composition of the influent and effluent from column tests is helpful to the evaluation. Geochemical evaluation may consist simply of a qualitative assessment of the potential for precipitate formation in the reactive cell based on site characterization and treatability data. Numerical geochemical codes may or may not be used depending on site objectives. Most available geochemical models are predictive and based on equilibrium codes, although inverse modeling codes can be used to backcalculate the mass of precipitating and dissolving compounds along a known flowpath. Inverse modeling codes, such as NETPATH, can be very useful for estimating net changes of material in the reactive cell. Furthermore, rates of inorganic chemical reactions can be calculated from the results when combined with measured flow velocities.

9.7 EMPLACEMENT OF THE BARRIER

Once the location, configuration, and dimensions of the permeable barrier have been designed, the best way to emplace the barrier in the ground needs to be determined. Chapter 6, Emplacement Techniques for Permeable Barrier Installation, describes the various methods available for installing the reactive cell and funnel walls (in case the barrier is a funnel-and-gate design). Because of the need to key the barrier into the aquitard, the depth of the aquitard is the primary parameter governing selection of the emplacement method. Geotechnical considerations, such as presence of rocks or highly consolidated sediments, may also affect the viability of the technique (for example, by affecting the drivability of sheet piles).

Emplacement of the reactive cell can be done by conventional trenching techniques using a backhoe or clamshell. Other techniques for emplacement of the reactive cell include the use of caissons, a continuous trencher, or a mandrel. For a funnel-and-gate configuration, the funnel walls are emplaced as impermeable barriers with techniques such as sheet piling or slurry walls.

9.8 MONITORING THE PERFORMANCE OF THE BARRIER

Once the emplacement of the barrier is complete, the barrier will have to be monitored for as long as the plume is present. The performance of the permeable barrier can be monitored with the following objectives:

- Evaluate adequate capture and treatment of plume and ensure acceptable downgradient water quality.

- Evaluate how well the barrier meets design objectives, such as residence time in the reactive cell.

- Evaluate the longevity of the barrier.

Monitoring may be conducted on a quarterly basis or on a different schedule as agreed to with the regulators. Experience with several permeable barrier installations to date suggests that the reactive medium is used up very slowly. Therefore, the required monitoring frequency is not expected to be very high. Other perform-ance monitoring measures, as described in Chapter 7, can provide early warnings of impending breakthrough. Hydrogeologic modeling can be very helpful, both for site engineers and regulators, as a tool for designing appropriate locations for monitoring wells. Column test data and geochemical evaluation (see Section 5.2) can provide helpful guidance in planning monitoring frequency.

Chapter 7 describes various monitoring techniques that may be used at the site manager's discretion to evaluate the barrier.

9.9 PERMEABLE BARRIER ECONOMICS

A cost-benefit approach should be used in evaluating the economic feasibility of a permeable barrier at a given site (see Chapter 8). Capital costs of a permeable barrier include the following:

- **Costs of the Reactive Medium**. These costs include two considerations — unit cost (price) of the medium and the total amount of medium required. The total amount of medium required in turn depends on the type and concentration of the contaminants, regulatory treatment criteria, the site groundwater velocity, and the contaminant distribution in the aquifer.

- **Costs of the Emplacement**. These costs depend on the depth of the aquitard, plume width, and geotechnical considerations (e.g., rocks or highly consolidated sediments).

- **Technology Licensing Costs**. Many of the reactive media and emplacement techniques are patented. Usually, the licensing costs of the emplacement techniques (e.g., sealed-joint sheet piling) are built into the quoted price, but licensing costs for the reactive medium will be an added cost for the site.

- **Disposal and Restoration Costs**. There may be costs associated with disposal of spoils generated during emplacement. The spoils may have to be disposed of as hazardous waste if the barrier is placed within the plume, incurring higher cost. The site surface may need to be restored to its original grade and condition.

The operating and maintenance (O&M) costs of the technology include the following:

- **Compliance Monitoring Costs**. These are the annual costs incurred for fulfilling regulatory requirements for monitoring breakthrough or bypassing of contaminants.

- **Additional Performance Monitoring Costs**. These may vary a lot from site to site depending on the objectives.

- **Periodic Maintenance Costs**. The reactive cell may have to be flushed or the reactive medium replaced periodically if precipitates build up to the point that reactivity or hydraulic performance is affected. Based on experience at existing sites, the incorporation of proper safety factors in the design may make it possible to keep the frequency of such maintenance as low as once in several years, if at all.

Any economic benefits of the permeable barrier application may be included in the evaluation as an offset, or reduction, to capital or O&M costs. Economic benefits may accrue, for example, from being able to put the property to more productive use because there is no need to install or operate an aboveground pump-and-treat system. Important intangible benefits, such as the risk reduction achieved, should also be considered.

CHAPTER 10

REFERENCES

Agrawal, A., and P.G. Tratnyek. 1994. "Abiotic Remediation of Nitro-Aromatic Groundwater Contaminants by Zero-Valent Iron." Proceedings of the 207th ACS National Meeting, San Diego, CA, *34*(1): 492-494.

Agrawal, A., and P. G. Tratnyek. 1996. "Reduction of Nitro Aromatic Compounds by Zero-Valent Iron Metal." *Environ. Sci. Technol. 30*(1): 153-160.

Amonette, J.E., J.E. Szecsody, H.T. Schaef, J.C. Templeton, Y.A. Gorby, and J.S. Fruchter. 1994. "Abiotic Reduction of Aquifer Materials by Dithionite: A Promising In-Situ Remediation Technology." In G.W. Gee and N.R. Wing (Eds.), *In-Situ Remediation: Scientific Basis for Current and Future Technologies*, pp. 851-881. Battelle Press, Columbus, OH.

Appleton, E.L. 1996. "A Nickel-Iron Wall Against Contaminated Groundwater." *Environmental Science and Technology 30*(12): 536A-539A.

Ballard, S. 1996. "The In-Situ Permeable Flow Sensor: A Ground-Water Flow Velocity Meter." *Groundwater 34*(2): 231-240.

Battelle. 1996. In-House Modeling Conducted at Battelle to Illustrate Application of Computerized Modeling Techniques for Permeable Barrier Design.

Battelle. 1997. September 1996 Monitoring Report for the Pilot Permeable Barrier at Moffett Federal Airfield: Draft Final. Prepared for Naval Facilities Engineering Service Center, Port Hueneme, CA, December 31.

Boronina, T., K.J. Klabunde, and G. Sergeev. 1995. "Destruction of Organohalides in Water Using Metal Particles: Carbon Tetrachloride/Water Reactions with Magnesium, Tin and Zinc." *Environ. Sci. Technol. 29*: 1511-1517.

Bostick, W.D., R.J. Jarabek, W.A. Slover, J.N. Fiedor, J. Farrell, and R. Heferich. 1996. *Zero-Valent Iron and Metal Oxides for the Removal of Soluble Regulated Metals in Contaminated Groundwater at a DOE Site, K/TSO-35-P*. Lockheed Martin Energy Systems, Inc., Oak Ridge, TN.

Breaux, B. 1996. Personal communication with Battelle from B. Breaux of Envirowall, Belle Chase, LA.

Burke, G.K. 1996. Personal communication with Battelle from G.K. Burke of Hayward Baker, Inc., Odneton, MD.

Burris, D.R., T.J. Campbell, and V.S. Manoranjan. 1995. "Sorption of Trichloroethylene and Tetrachloroethylene in a Batch Reactive Metallic Iron-Water System." *Environ. Sci. Technol. 29*(11): 2850-2855.

Cantrell, K.J., and D.I. Kaplan. 1996. "Zero-Valent Iron Colloid Emplacement in Sand Columns." *J. Environ. Eng.* (In press).

Cantrell, K.J., D.I. Kaplan, and T.J. Gilmore. 1997. "Injection of Colloidal Fe^0 Particles in Sand Columns with Shearthinning Fluids." *J. Environ. Eng.* (Submitted).

Cavalli, N.J. 1992. "Composite Barrier Slurry Wall." In D.B. Paul, R.R. Davidson, and N.J. Cavalli (Eds.), *Slurry Walls: Design, Construction and Quality Control, ASTM STP 1129.* American Society for Testing and Materials, Philadelphia, PA.

Chapelle, F.H. 1993. *Ground-Water Microbiology & Geochemistry.* John Wiley & Sons, New York, NY.

Day, S. 1996. Personal communication with S. Day of Geo-Con, Inc., Denver, CO.

Delta Research Corporation. 1996. *RACER 3.2 for the U.S. Air Force, User Manual.*

Deng, B., T. Campbell, and D.R. Burns. 1996. "Hydrocarbon Formation in Metallic Iron/Water Systems." Accepted for publication in *Environmental Science and Technology.*

EPA, see U.S. Environmental Protection Agency.

ETI. 1995. Marketing literature from Envirometal Technologies, Inc., Guelph, Ontario.

ETI. 1996. Personal communication from J. Vogan, Envirometal Technologies, Inc., Guelph, Ontario.

ETI. 1997. Personal communication from J. Vogan and S. O'Hannesin, Envirometal Technologies, Inc., Guelph, Ontario.

Fetter, C.W. 1994. *Applied Hydrogeology.* 3rd ed. Merrill Publishing Company, Columbus, OH.

Focht, R.M. 1994. Bench-Scale Treatability Testing to Evaluate the Applicability of Metallic Iron for Above-Ground Remediation of 1,2,3-Trichloropropane Contaminated Groundwater. M.Sc. Thesis, Department of Earth Sciences, University of Waterloo, Waterloo, Ontario, pp. 58.

Focht, R.M., J.L. Vogan, and S.F. O'Hannesin. 1997. Hydraulic Studies of In-Situ Permeable Reactive Barriers. Unpublished paper, University of Waterloo.

Gillham, R.W. 1993. Cleaning Halogenated Contaminants from Groundwater. U.S. Patent No. 5,266,213, Nov. 30.

Gillham, R.W. 1996. "In Situ Treatment of Groundwater: Metal-Enhanced Degradation of Chlorinated Organic Contaminants." In M.M. Aral (Ed.), *Advances in Groundwater Pollution Control and Remediation,* pp. 249-274. Kluwer Academic Publishers.

Gillham, R.W. 1997. Personal communication with Battelle from R.W. Gillham and John Vogan of Envirometal Technologies, Inc., Guelph, Ontario.

Gillham, R.W., and S.F. O'Hannesin. 1992. "Metal-Catalyzed Abiotic Degradation of Halogenated Organic Compounds." *IAH Conference: Modern Trends in Hydrogeology.* Hamilton, Ontario, May 10-13, pp. 94-103.

Gillham, R.W., and S.F. O'Hannesin. 1994. "Enhanced Degradation of Halogenated Aliphatics by Zero-Valent Iron." *Ground Water 32*: 958-967.

Gillham, R.W., S.F. O'Hannesin, and W.S. Orth. 1993. "Metal Enhanced Abiotic Degradation of Halogenated Aliphatics: Laboratory Tests and Field Trials." Paper presented on March 9-11, HazMat Central Conference, Chicago, IL.

Gorsky, G. 1996. Personal communication with Battelle from G. Gorsky of the Keller Environmental Group, Chicago, IL.

Groundwater Control, Inc. 1996. "Here's How It Works...Groundwater Control, Inc. One-Pass Deep Trencher." Soil and Groundwater Cleanup Informational Brochure. Jacksonville, FL.

Hadley, P. 1996. Personal communication from P. Hadley, California Department of Toxic Substances Control, September 19.

Hardy, L.I., and R.W. Gillham. 1996. "Formation of Hydrocarbons from the Reduction of Aqueous CO_2 by Zero-Valent Iron." *Environ. Sci. Technol. 30*(1): 57-65.

Holser, R.A., S.C. McCutcheon, and N.L. Wolfe. 1995. "Mass Transfer Effects on the Dehalogenation of Trichloroethene by Iron/Pyrite Mistures." *Extended Abstracts from the 209th ACS National Meeting, Anaheim, Cal., 35*(1), 788-791. Anaheim, CA. Division of Environmental Chemistry, American Chemical Society, Washington, DC.

Jeffers, P.M., L.M. Ward, L.M. Woytowitch, and N.L. Wolfe. 1989. "Homogeneous Hydrolysis Rate Constants for Selected Chlorinated Methanes, Ethanes, Ethenes and Propanes." *Environ. Sci. Technol. 23*(8): 965-969.

Kaplan, D.I., K.J. Cantrell, T.W. Wietsma, and M.A. Potter. 1996. "Formation of a Chemical Barrier with Zero-Valent Iron Colloids for Groundwater Remediation." *J. Environ. Qual. 25*: 1086-1094.

Kearl, P., N. Korte, M. Stites, and J. Baker. 1994. "Field Comparison of Micropurging vs. Traditional Ground Water Sampling." *Ground Water Monitoring and Remediation 14*(4): 183-190.

Kilfe, H. 1996. Personal communication from H. Kilfe, Water Resources Control Engineer, California Regional Quality Control Board.

Korte, N.E., L. Liang, and J. Clausen. 1995. "The Use of Palladized Iron as a Means of Treating Chlorinated Contaminants." *Emerging Technologies in Hazardous Waste Management VII, Extended Abstracts for the Special Symposium*, pp. 42-45. Atlanta, GA.

Kriegman-King, M.R., and M. Reinhard. 1991. "Reduction of Hexachloroethane and Carbon Tetrachloride at Surfaces of Biotite, Vermiculite, Pyrite, and Marcasite." In R. A. Baker (Ed.), *Organic Substances and Sediments in Water*, Vol. 2. Lewis Publishers, Chelsea, MI.

Kriegman-King, M.R., and M. Reinhard. 1994. "Transformation of Carbon Tetrachloride by Pyrite in Aqueous Solution." *Environ. Sci. Technol. 28*: 692-700.

Lipczynska-Kochany, E., S. Harms, R. Milburn, G. Sprah, and N. Nadarajah. 1994. "Degradation of Carbon Tetrachloride in the Presence of Iron and Sulphur Containing Compounds." *Chemosphere 29*: 1477-1489.

Mackenzie, P.D., S. Sunita, G.R. Eykholt, D.P. Horney, J.J. Salvo, and T.M. Sivavec. 1995. "Pilot-Scale Demonstration of Reductive Dechlorination of Chlorinated Ethenes by Iron Metal." Presented at the 209th ACS National Meeting, Anaheim, CA, April 2-6.

Matheson, L.J. and P.G. Tratnyek. 1994. "Reductive Dehalogenation of Chlorinated Methanes by Iron Metal." *Environ. Sci. Technol. 28*: 2045-2053.

Muftikian, R., Q. Fernando, and N. Korte. 1995. "A Method for the Rapid Dechlorination of Low Molecular Weight Chlorinated Hydrocarbons in Water." *Water Res. 29*: 2434.

Myller, B. 1996. Personal communication to Battelle from B. Myller of Dames and Moore.

National Research Council. 1994. *Alternatives for Ground Water Clean Up*. National Academy Press, Washington, DC.

O'Hannesin, S.F. 1993. A Field Demonstration of a Permeable Reaction Wall for the In Situ Abiotic Degradation of Halogenated Aliphatic Organic Compounds. Unpublished M.S. thesis, University of Waterloo, Ontario, Canada.

Orth, R.G., and D.E. McKenzie. 1995. "Reductive Dechlorination of Chlorinated Alkanes and Alkenes by Iron Metal and Metal Mixtures." *Extended Abstracts from the Special Symposium Emerging Technologies in Hazardous Waste Management VII*, p. 50. American Chemical Society, Atlanta, GA.

Orth, W.S., and R.W. Gillham. 1995. "Chloride and Carbon Mass Balances for Iron-Enhanced Degradation of Trichloroethene." Presented at the 209th ACS National Meeting, Anaheim, CA, April 2-6.

Orth, W.S., and R.W. Gillham. 1996. "Dechlorination of Trichloroethene in Aqueous Solution Using Fe(0)." *Environ. Sci. Technol. 30*(1): 66-71.

Owaidat, L. 1996. Personal communication to Battelle from L. Owaidat of Geo. Con., Inc., Rancho Cordova, CA.

PRC. See PRC Environmental Management, Inc.

PRC Environmental Management, Inc. 1996. *Naval Air Station Moffett Field, California, Iron Curtain Area Groundwater Flow Model*. PRC, June.

Puls, R.W., R.M. Powell, and C.J. Paul. 1995. "In Situ Remediation of Ground Water Contaminated with Chromate and Chlorinated Solvents Using Zero-Valent Iron: A Field Study." *Extended Abstracts from the 209th ACS National Meeting Anaheim, CA, 35*(1): 788-791. Anaheim, CA. Division of Environmental Chemistry, American Chemical Society, Washington, DC.

Reardon, E.J. 1995. "Anaerobic Corrosion of Granular Iron: Measurement and Interpretation of Hydrogen Evolution Rates." *Environ. Sci. Technol. 29*(12): 2936-2945.

Reynolds, G.W., J.T. Hoff, and R.W. Gillham. 1990. "Sampling Bias Caused by Materials Used to Monitor Halocarbons in Groundwater." *Environ. Sci. Technol. 24*(1): 135-142.

Roberts, A.L., L.A. Totten, W.A. Arnold, D.R. Burris, and T.J. Campbell. 1996. "Reductive Elimination of Chlorinated Ethylenes by Zero-Valent Metals." *Environ. Sci. Technol.*

Schmithorst, B. 1996. Personal communication with Battelle from W. Schmithorst of Parsons Engineering Science.

Schreier, C.G., and M. Reinhard. 1994. "Transformation of Chlorinated Organic Compounds by Iron and Manganese Powders in Buffered Water and in Landfill Leachate." *Chemosphere 29*: 1743-1753.

Senzaki, T. 1988. "Removal of Chlorinated Organic Compounds from Wastewater by Reduction Process: III. Treatment of Tetrachloroethane with Iron Powder II." *Kogyo Yosui 391:* 29-35 (in Japanese).

Senzaki, T., and Y. Kumangai. 1988a. "Removal of Chlorinated Organic Compounds from Wastewater by Reduction Process: Treatment of 1,1,2,2-Tetrachloroethane with Iron Powder." *Kogyo Yosui 357:* 2-7 (in Japanese).

Senzaki, T., and Y. Kumangai. 1988b. "Removal of Chlorinated Organic Compounds from Wastewater by Reduction Process: II. Treatment of Tetrachloroethane with Iron Powder." *Kogyo Yosui 369:* 19-25 (in Japanese).

Sivavec, T. 1996. Personal communication from T. Sivavec, General Electric Corporate Research and Development, Schenectady, NY.

Sivavec, T. 1997. Personal communication from T. Sivavec, General Electric Corporate Research and Development, Schenectady, NY.

Sivavec, T.M., and D.P. Horney. 1995. "Reductive Dechlorination of Chlorinated Ethenes by Iron Metal." Presented at the 209th ACS National Meeting, Anaheim, CA. April 2-6.

Sivavec, T.M., D.P. Horney, and S.S. Baghel. 1995. "Reductive Dechlorination of Chlorinated Ethenes by Iron Metal and Iron Sulfide Minerals." *Emerging Technologies in Hazardous Waste Management VII,* pp. 42-45. Extended Abstracts for the Special Symposium, Atlanta, GA.

Sivavec, T.M., P.D. Mackenzie, and D.P. Horney. 1997. "Effect of Site Groundwater on Reactivity of Bimetallic Media: Deactivation of Nickel-Plated Granular Iron." Pre-print paper, ACS National Meeting, American Chemical Society, Division of Environmental Chemistry, *37*(1).

Starr, R.C., and J.A. Cherry. 1994. "In Situ Remediation of Contaminated Ground Water: The Funnel-and-Gate System." *Groundwater 32*(3): 465-476.

Sweeny, K.H. 1981a. "The Reductive Treatment of Industrial Wastewaters: I. Process Description." In G. F. Bennett (Ed.), *American Institute of Chemical Engineers, Symposium Series, Water-1980, 77* (209): 67-71.

Sweeny, K.H. 1981b. "The Reductive Treatment of Industrial Wastewaters: II. Process Description." In G. F. Bennett. (Ed.), *American Institute of Chemical Engineers, Symposium Series, Water-1980, 77* (209): 72-88.

Sweeny, K.H. 1983. Treatment of Reducible Halohydrocarbons Containing Aqueous Stream. U.S. Patent 4,382,865.

Sweeny, K.H., and J.R. Fischer. 1972. Reductive Degradation of Halogenated Pesticides. U.S. Patent No. 3,640,821.

Sweeny, K.H., and J.R. Fischer. 1973. Decomposition of Halogenated Pesticides. U.S. Patent 3,737,384.

Turner, M. 1996. Personal communication from M. Turner, New Jersey Department of Environmental Protection, September 24.

U.S. Environmental Protection Agency. 1989. *Pocket Guide for the Preparation of Quality Assurance Project Plans*. EPA/600/9-89/087.

U.S. Environmental Protection Agency. 1991. EPA Method 624, *Purgeables*. EPA SW-846, July.

U.S. Environmental Protection Agency. 1994. EPA Method 8240, *Volatile Organic Compounds by Gas Chromatography/Mass Spectrometry (GC/MS)*. EPA SW-846, Update II, September.

U.S. Environmental Protection Agency. 1995. *In Situ Remediation Technology Status Report: Treatment Walls*. EPA 542-K-94-004. U.S. EPA, Office of Solid Waste and Emergency Response, April.

U.S. Environmental Protection Agency. 1996. *A Citizen's Guide to Treatment Walls*. EPA 542-F-96-016. September, 1996.

U.S. EPA, see U.S. Environmental Protection Agency.

Wagner, K., K. Boyer, R. Claff, M. Evans, S. Henry, V. Hodge, S. Mahmud, D. Sarno, E. Scopino, and P. Spooner. 1986. *Remedial Action Technology for Waste Disposal Sites*. Noyes Data Corporation, Park Ridge, NJ.

Warner, S.D., C.L. Yamane, J.D. Gallinati, and D.A. Hankins. 1996. "Considerations for Monitoring Permeable Groundwater Treatment Walls." Accepted for publication by *ASCE Journal of Environ. Eng.*

ADDITIONAL SITE CHARACTERIZATION AND MONITORING ISSUES

A.1 SITE PROFILE SHEET

An example site profile sheet that can be used to characterize the hydrogeology and geochemistry of the site and evaluate its suitability for permeable barrier application appears on the following two pages. As much of the information requested on this profile sheet as possible should be provided using existing site documents. Based on this information, a preliminary decision can be made regarding the suitability of the site for permeable barrier application. Data gaps identified during this information-gathering exercise can be addressed by conducting additional site characterization to aid in the design of the barrier.

A.2 GROUNDWATER FLOW SYSTEM CHARACTERISTICS IMPORTANT FOR PERMEABLE BARRIER DESIGN

A.2.1 Background Information

A preliminary characterization of the site geology is necessary to identify formation characteristics that may affect groundwater flow, contaminant movement, and permeable wall design. A search for background geologic information should be completed as part of this characterization. It is anticipated that some of this information would be available from previous site characterization studies conducted at the site. This could include a literature search, and a review of maps and reports. The U.S. Geological Survey, state geological surveys, geological societies, agricultural organizations (such as the U.S. Department of Agriculture), environmental protection agencies, regional and local development authorities, and local universities are likely sources of both regional and local geologic information. Aerial photographs can be used to evaluate surface water features, manmade drainage networks, irrigation systems, and structural features that may affect groundwater movement. Water well construction and abandonment records, oil and gas well records, surface mining permits, and geotechnical borings should be identified and reviewed for information on subsurface geology. These records may be available from state geological surveys and local or county health departments. A

SITE PROFILE SUMMARY

Site Location   _____

Site Name/Number _____

Hydrogeologic Parameters

Type of aquitard (clay/bedrock) _____

Aquifer soil texture (sand/silt/clay/other) _____

Depth to water table   _____ ft bgs   Groundwater flow velocity _____ ft/day

Depth to aquitard   _____ ft bgs   Thickness of aquitard   _____ ft

Groundwater flow direction (attach water-level map if available) _____

_____

Distance to downgradient property boundary from the edge of the plume _____

_____

Distance of the nearest receptor (e.g., drinking water well) from the edge of the

plume _____

List any underground obstacles in the region of the plume (e.g., underground

utilities) _____

_____

_____

List any aboveground obstacles in the region of the plume (e.g., buildings) _____

_____

_____

_____

Is a pump-and-treat system currently operating or planned? _____

_____

FIGURE A-1. Example site profile sheet

SITE PROFILE SUMMARY (continued)   page 2

Contaminants and Groundwater Geochemistry

List organic and/or inorganic contaminants detected in groundwater and their
maximum concentrations (e.g., trichloroethylene, 2.0 mg/L):

_____	_____
_____	_____
_____	_____
_____	_____
_____	_____
_____	_____
_____	_____
_____	_____

Other Groundwater Parameters and Maximum Concentrations

Alkalinity	_____ mg/L	Total dissolved solids	_____ mg/L
Dissolved oxygen	_____ mg/L	Total organic carbon	_____ mg/L

Calcium	_____ mg/L	Chloride	_____ mg/L
Iron	_____ mg/L	Nitrate	_____ mg/L
Magnesium	_____ mg/L	Sulfate	_____ mg/L
Manganese	_____ mg/L		

Attach if Available:

- Site map
- Plume map
- Water-level map

- Monitoring well network map
- Geologic cross-sections (or well logs)
- Underground utilities map

FIGURE A-1. Example site profile sheet (cont'd)

site walkover should also be completed. At this time, surface features identified in the geologic information, maps, and photographs can be verified. Existing wells, surface structures, and construction constraints can be identified and potential locations for the permeable barrier can be evaluated. Areas of exposed geologic strata at and adjacent to the site should also be inspected if accessible. Finally, general information on the type and extent of contamination should be compiled.

The geologic background and site information should be assembled and a preliminary conceptual model of the subsurface geologic features constructed. This model should have general information on the site lithology, various aquifer layers and confining units, contaminant plume configuration, and factors such as precipitation, design constraints due to surface and subsurface features (e.g., buildings, tunnels, boundaries), and regulatory constraints. A conceptualization of the sedimentary facies variations also should be developed. These facies variations have a significant impact on aquifer heterogeneity, which may be the most important control on the groundwater flow system. The model can be used as a basis for further exploration and development of the site-specific data for permeable barrier installation.

A.2.2 Hydrostratigraphic Framework

The hydrostratigraphic framework involves collection and synthesis of site-specific geologic and hydrologic data to construct a conceptual understanding of groundwater flow at the site. This may include the information collected from previous reports and the new information collected specifically for the permeable barrier installation. The most significant data to be collected include variations in the depth, thickness and water levels of different hydrostratigraphic units (HSUs). This is achieved by conventional drilling and sampling at several locations or by some innovative technique, such as the cone penetrometer test (CPT) or the Geoprobe methods. Monitoring wells, soil borings, and production well borings can be installed at the site to characterize the subsurface geology. The number and location of boreholes and samples required for site characterization will depend mainly on the site heterogeneity and availability of preexisting data, and should be based on the scientific judgment of the on-site hydrogeologist. At relatively homogeneous sites, only a few boreholes are needed to characterize the site adequately. However, at sites with heterogeneous sediments, a large number of boreholes are needed before a reliable picture of the subsurface features can be developed. Water levels should be measured from all the monitoring wells and piezometers installed in the vicinity of the site. These water-level measurements should be repeated several times over the year to estimate the seasonal variations in water levels at the site. Alternatively, continuous water-level recorders can be installed in one or more monitoring points to record water-level variations.

Some intact formation samples (cores, split-spoons, thin-walled tubes) should be collected to provide a field description of the geologic conditions and to identify or estimate the hydrogeologic properties of the aquifer. Samples can be collected to chemically characterize the sediments and to measure the physical and geochemical properties of the aquifer. Groundwater samples can be collected to characterize the nature and extent of contamination and general groundwater quality. Formation samples can be analyzed to measure physical properties (grain size, mineralogy, lithology, texture, etc.) and hydrogeologic properties (porosity, permeability). Samples should be described and logged in the field and, if appropriate, submitted to a soils laboratory for analysis. Laboratory analysis of hydraulic conductivity (K) and porosity can be used in the development of design requirements for the permeable cells because the residence times, flow velocity, and discharge are based on these parameters. Other laboratory analyses can be completed to evaluate concentrations of adsorbed contaminants and to evaluate geochemical properties, such as organic carbon content of the aquifer material.

Once all of the field and laboratory data have been obtained, site-specific geologic cross-sections should be prepared to evaluate the lithologic variations at the site. In addition, water-level data from shallow and deep HSUs should be plotted on the cross-section to evaluate the vertical hydraulic gradients across the adjacent aquifers. The water levels also should be plotted as contour maps that can be used to determine the groundwater flow directions, hydraulic gradients, and velocities at the site. These maps may help in determining the effects of any significant sources and sinks, such as wells, streams, tunnels, etc., on the groundwater flow system. Finally, the water-level data collected at different times should be plotted in water-versus-time and water-level contour maps. These plots are useful in determining seasonal variations in water levels and flow directions and incorporating their effects into the permeable wall design.

As an example, the site characterization for the pilot-scale permeable barrier at Moffett Federal Airfield and the commercial full-scale installation in Sunnyvale, California and other sites involved the data from several boreholes and numerous CPTs. In addition, information from preexisting pumping tests and water-level data were also used. Detailed site-specific cross-sections of subsurface lithologic variations were constructed. These lithologic cross-sections were correlated with the base-wide maps of subsurface sand channel deposits that act as preferential pathways for most of the groundwater flow and contaminant transport. The pilot-scale reactive cell was placed across one of these channels, and the funnel walls were placed across the finer grained interchannel deposits.

A.2.3 Hydrologic Parameter Estimation

The aquifer parameters that need to be collected or determined as part of a complete site characterization include the hydraulic conductivity of the aquifer materials, in both the horizontal and vertical directions, the porosity of the geologic media, the lateral and vertical components of hydraulic gradient of the flow field, and the ambient groundwater flow directions and velocities.

A.2.3.1 Hydraulic Conductivity (K). K is the measure of an aquifer's ability to transmit water and is expressed as the rate at which water can move through a unit thickness permeable medium. K is perhaps the most important aquifer parameter governing fluid flow in the subsurface. The velocity of groundwater movement and dissolved contaminant migration is directly related to the K of the saturated zone. In addition, subsurface variations in K directly influence contaminant fate and transport by providing preferential pathways for contaminant migration. Estimates of K are used to determine flow velocities and travel times for contaminants and groundwater. At heterogeneous sites, most of the groundwater flow and contaminant transport in the aquifers may be restricted to high K zones. It is important to delineate these preferential pathways by a combination of geologic and hydrologic characterization so that the permeable barriers can be located across these zones.

The most common methods used to quantify K are single- and multiple-well aquifer tests and slug tests. A drawback to these methods is that the resulting K values represent "average" values of conductivity over the length of the well screen used in the test well. In addition, it is important to match the scale of the aquifer test with the scale of the study area of interest. At sites with significant vertical heterogeneities or anisotropy, it would be useful to estimate vertical K. These data can be important in estimating the extent of overflow, underflow, and cross-formational flow. Vertical K can be determined from pumping tests or from laboratory testing of sediment cores. In addition to K, some aquifer tests provide an estimate of transmissivity (τ). This parameter is simply a multiple of K and aquifer thickness and provides a site-specific measure of the ability of a particular aquifer to transmit water.

A.2.3.2 Pumping Tests. Pumping tests involve the pumping of a test well and observation of the pumping stress in the surrounding area through the measurement of hydraulic head in the pumped well and several surrounding observation wells. A complete description of the theory and application of pumping tests can be found in Domenico and Schwartz (1990) and Fetter (1994). A complete description of pumping tests and the various methods that can be used in the analysis of data collected during a pumping test is provided in Kruseman and de Ridder (1991).

Pumping tests generally give the most reliable information on K, but they may be difficult to conduct in contaminated areas because the water produced during the test generally must be contained and treated. In addition, a minimum 4-inch-diameter well is generally required to conduct pumping tests in highly transmissive aquifers because the 2-inch submersible pumps currently available are not capable of producing flowrates large enough to induce significant drawdown. In areas with fairly uniform, or homogeneous, aquifer materials, pumping tests may be completed in uncontaminated areas, and the results can be used to estimate K in the contaminated area. Pumping tests are relatively expensive and time-consuming to conduct. Therefore, at highly heterogeneous sites, it may be difficult to conduct a sufficient number of tests to estimate the K variations in all the hydrostratigraphic units of interest.

It is important that pumping tests be conducted in wells that are capable of yielding the amount of water being pumped during the tests, and that flow into the well not be constricted. If these conditions are not met, the pumping test is testing the transmissive capacity of the well screen in the test well rather than the ability of the aquifer to transmit water. In addition, it is important that the pumping test last for a sufficient length of time. This is particularly true in unconfined aquifers, in which a delayed aquifer response may occur. It is standard practice to conduct 72-hour pumping tests in unconfined aquifers.

A.2.3.3 Slug Tests. Slug withdrawal or injection tests are a commonly used alternative to pumping tests. A slug test consists of the insertion or removal of a "slug" or known volume of water, or the displacement of water by a solid object. The displaced water causes a stress on the aquifer formation, which is monitored through the change and recovery of hydraulic head or water level. One commonly cited limitation to slug testing is that the method generally gives K information only for the area immediately surrounding the test well. Slug tests do, however, have two distinct advantages over pumping tests; they can be conducted in small-diameter wells, and they produce no contaminated water that must be treated and/or disposed of. If slug tests are to be used as part of a site characterization effort to determine the K distribution in an aquifer, it is important that multiple slug tests be performed. It is advisable to not rely on data from one slug test in a test well. The tests should be performed with replicates and in as many test or monitoring wells as feasible. One of the biggest advantages of slug tests over the pumping tests is that a large number of tests can be conducted in the amount of time and cost it takes for one pumping test. Therefore, slug tests can be used to estimate the spatial variations in K at heterogeneous sites. A description of the theory and application of slug testing is provided in Fetter (1994), and a complete description of the analysis of slug test data is provided in Kruseman and de Ridder (1991).

A.2.3.4 Porosity. The porosity of an aquifer material is the percentage of the rock or soil/sediment that consists of void space. The porosity of a sample of aquifer material can be determined relatively easily in the laboratory. This is done by drying the sample to remove any moisture clinging to the surfaces in the sample but not water that is hydrated as a part of certain minerals. The dried sample is then submerged in a known volume of water and allowed to remain in a sealed chamber until it is saturated. The volume of voids is equal to the original water volume less the volume in the chamber after the saturated sample is removed. This method excludes pores not large enough to contain water molecules and those which are not interconnected.

The total porosity can be estimated from the equation

$$n = 100[1 - (P_b/P_d)]$$

where n =   the porosity (percentage)
          P_b =   the bulk density of the aquifer material (g/cm^3)
          P_d =   the particle density of the aquifer material (g/cm^3)

The bulk density of the aquifer material is the mass of the sample after drying divided by the original sample volume. The particle density is the oven-dried mass divided by the volume of the mineral matter in the sample as determined by the water-displacement test (Fetter, 1994).

Table A-1 lists general ranges of porosity that can be expected for some typical sediments.

TABLE A-1. Porosity ranges for sediments

Well-sorted sand or gravel	25-50%
Sand and gravel, mixed	20-35%
Glacial till	10-20%
Silt	35-50%
Clay	33-60%

A.2.3.5 Hydraulic Gradient. The hydraulic gradient is defined as the change in hydraulic head with a change in distance in the direction which yields a maximum rate of decrease in head. The hydraulic gradient can be determined from a water-table or potentiometric surface map constructed using water-level measurements taken at the site during a specific time. It is important to estimate values of both the lateral and vertical hydraulic gradients of the site. The vertical gradients are useful in evaluating potential for underflow or overflow and flow between adjacent aquifers. The vertical hydraulic gradients may be determined by comparing water levels in multiple well clusters with individual points screened at different vertical depths.

A.2.3.6 Groundwater Flow Direction and Velocity. Groundwater flow directions may also be determined from a water-table or potentiometric surface map based on water-level measurements made at the site. Groundwater flows perpendicular to equipotential lines are expressed on a map as contours of water-table or potentiometric surface elevation.

The average linear groundwater flow velocity can be determined from the K of the aquifer materials, the hydraulic gradient of the flow system, and the effective porosity of the aquifer materials. The average linear groundwater flow velocity can be calculated by

$$V_x = \frac{K(dh/dl)}{n_e}$$

where　V_x　=　the average linear groundwater flow velocity
　　　　K　=　the hydraulic conductivity of the aquifer material (L/t)
　　　dh/dl　=　the hydraulic gradient
　　　　n_e　=　the effective porosity

Groundwater flow directions and velocities may also be calculated using a three-point problem approach, which enables the calculation of velocities in three-space dimensions from hydraulic head measurements as described by Pinder and Abriola (1982). Recently, some direction measurement techniques have been developed using in situ flow sensors (Ballard, 1996) to estimate flow velocity and flow directions. The final design of the permeable barrier should incorporate the effect of maximum variation in flow directions to avoid future situations where the plume may bypass the barrier.

A.3 MICROBIAL INFLUENCES IN PERMEABLE BARRIER APPLICATIONS

Redox-sensitive elements, such as N and S, are closely associated with biological processes, and as a result, their concentrations in groundwater may not agree with predictions based on thermodynamics. Based on thermodynamic calculations, metallic iron would be thought to completely reduce nitrate and sulfate to their lowest oxidation states, ammonia and sulfide, respectively. However, in reactions for which biological mechanisms play a key role, the approach to equilibrium depends on the presence of enzymes from bacteria and other microorganisms. Furthermore, it is important to consider that these microorganisms require oxidizable organic matter as an energy source. Catalysis provided by enzyme systems is important in redox reactions involving N and S because activation energy barriers may be so high as to substantially inhibit progress of these reactions.

In aerobic groundwater, nitrogen is present predominantly as dissolved NO_3^-, with lesser amounts of N_2. Because neither species takes part in mineral formation reactions, their concentrations tend to remain stable under natural conditions. Under anaerobic conditions, bacteria can reduce NO_3^- to N_2 by the process of denitrification, as shown in Equation A-1.

$$5CH_2O + 4NO_3^- \rightarrow 2N_2 + 4HCO_3^- + CO_2 + 3H_2O \qquad (A-1)$$

In Equation A-1, CH_2O represents organic matter in the form of carbohydrate. Continued reduction would lead to removal of N_2 and formation of ammonia by the process of fixation, needed to form bacterial protein.

Another concern is that any chemical or biochemical process that increases pH will affect the balance of bicarbonate and carbonate ions, and this may cause precipitation of dissolved Ca^{2+} as calcite, as described by Equations A-2 and A-3. Precipitation of calcite is a potential concern with regard to clogging of the reactive cell.

$$2HCO_3^- + Ca^{2+} \rightarrow CaCO_3 + CO_2 + H_2O \qquad (A-2)$$

$$CO_3^{2-} + Ca^{2+} \rightarrow CaCO_3 \qquad (A-3)$$

Another potential concern is precipitation of siderite ($FeCO_3$), which reacts with bicarbonate and carbonate in the same stoichiometry as calcite (similar to calcium in Equations A-2 and A-3).

In most natural anaerobic waters, dissolved sulfide is produced by bacterial reduction of sulfate. This general reaction involves oxidation of organic matter, such as carbohydrate, as shown in Equation A-4.

$$2CH_2O + SO_4^{2-} \rightarrow H_2S + 2HCO_3^-$$ (A-4)

Bacteria such as *Desulfovibrio* are obligate anaerobes that may not become acclimated within a reactive cell. This explains why sulfate levels can remain high and sulfide (or bisulfide) levels may be very low in column experiments or permeable barriers containing metallic iron. This is fortunate, because sulfate in groundwater can be as high as a few hundred mg/L. If sulfate were reduced to sulfide, substantial amounts of mineral precipitates such as FeS could form and potentially lead to reduced cell permeability.

Some studies on aboveground treatment of groundwater have also reported iron-related potential for biological activity. For example, biofouling is a term applied to microbiological processes that result in oxidation of ferrous ions [Fe(II)] and subsequent precipitation of ferric [Fe(III)] hydroxides. Iron-related biofouling has been attributed to various types of clogging problems in groundwater treatment systems (Chapelle, 1993), and there has been speculation that such problems could be encountered in permeable barriers.

The predominant concern expressed relates to iron bacteria populations that have been observed in aquifers at sites using pump-and-treat systems, and the related plugging of well screens and other treatment equipment. However, geochemical conditions and bacterial populations in an in situ permeable barrier of reactive iron may not be the same as those of standard groundwater pumping/monitoring wells. In a well screen, relatively reduced groundwater containing dissolved iron enters an oxygenated environment in the wellbore, creating conditions where iron-oxidizing bacteria can cause fouling problems. In a permeable barrier, groundwater becomes even more reducing as it moves through the iron, which represents a relatively "hostile" environment for bacterial growth, as discussed below.

To date, no sliming or other visual evidence of microbial activity has been observed in more than 50 different laboratory-scale treatability studies using groundwater exhibiting a wide range of inorganic and organic chemistry (ETI, 1997). There is also a significant question as to the relevance of microbial analyses conducted on column samples to any in situ microbial activity which might occur, as the column tests are done at room temperatures using groundwater that has been at least partially degassed and exposed to the atmosphere due to sample collection and transport. Tests using "sterile" and "unsterile" influent solutions of the same groundwater have resulted in the same VOC degradation rates (Gillham and O'Hannesin, 1994), indicating that microbial effects on VOC degradation at the laboratory scale are insignificant.

The effects of microbial activity have also been examined at the field scale. While declines in sulfate concentrations at some sites have indicated the presence of sulfate-reducing bacteria, cores of the permeable barrier at the Borden test site, collected two years after the barrier's installation, showed no significant population of iron-oxidizing microbes, and only low numbers of sulfate-reducers (Matheson and Tratnyek, 1993 and 1994). No evidence of microbial fouling or performance decrease has been observed at the Borden site, installed in 1991. Phospholipid-fatty acid analysis of groundwater from an aboveground test reactor at an industrial facility in California showed no enhanced microbial population in the reactive media relative to background groundwater samples. Also, an aboveground reactor has been operating at temperatures of 4 to 15°C since October 1994 in New Jersey with no indications of media fouling (ETI, 1997).

The most detailed sampling of an in situ installation was completed at a pilot-scale permeable barrier in upstate New York (ETI, 1997). Data on microbial biomass and composition were collected from wells in the reactive cell (iron medium), in upgradient and downgradient pea gravel zones, and in upgradient and downgradient aquifer monitoring wells for a 6-month period following system construction. The "background" microbial community in the aquifer at this site appeared to be disrupted during construction, and then re-established itself in the new environment created by the gate.

There was no evidence of significant microbial growth in the upgradient pea gravel or in the reactive cell; the microbial populations in the upgradient pea gravel, reactive cell, and surrounding aquifer background appeared similar in size and composition. The microbial biomass in the downgradient pea gravel and downgradient aquifer was approximately 10 times greater than the microbial biomass in the aquifer background, upgradient pea gravel, and reactive cell, and of different composition. The difference in microbial population size and composition could possibly be explained by the changes in the geochemistry created by the permeable barrier on the downgradient side of the reactive cell, notably the production of hydrogen gas from the iron that supports the activity of many obligate anaerobic bacteria such as sulfate-reducing bacteria. Although the biomass increased on the downgradient side of the barrier, the amount of biomass was not particularly large relative to microbial biomass in surface or agricultural soils. The limited availability of nutrients (i.e., hydrogen) will likely limit any substantial increase in biomass in the downgradient pea gravel and downgradient aquifer zones past this population size.

In summary, microbial activity appears to have had little effect on the performance of the reactive cell in both laboratory and field tests conducted so far. Therefore, current knowledge indicates no significant impacts on long-term performance of this technology due to microbial activity.

A.4 APPENDIX A REFERENCES

Ballard, S. 1996. "The In-Situ Permeable Flow Sensor: A Ground-Water Flow Velocity Meter." *Groundwater 34*(2): 231-240.

Chapelle, F.H. 1993. *Ground-Water Microbiology & Geochemistry*. John Wiley & Sons, New York, NY.

Domenico, P.A., and F.W. Schwartz. 1990. *Physical and Chemical Hydrogeology*. John Wiley & Sons, Inc., New York, NY.

ETI. 1997. Personal communication from J. Vogan and S. O'Hannesin, Envirometal Technologies, Inc., Guelph, Ontario.

Fetter, C.W. 1994. *Applied Hydrogeology*. 3rd ed. Merrill Publishing Company, Columbus, OH.

Gillham, R.W., and S.F. O'Hannesin. 1994. "Enhanced Degradation of Halogenated Aliphatics by Zero-Valent Iron." *Ground Water 32*: 958-967.

Kruseman, G.P., and N.A. de Ridder. 1991. *Analysis and Evaluation of Pumping Test Data*, 2nd ed. (Completely Revised). International Institute for Land Reclamation and Improvement, Wageningen, The Netherlands, Publication 47.

Matheson, L.J., and P.G. Tratnyek. 1993. "Processes Affecting Reductive Dechlorination of Chlorinated Solvents by Zero-Valent Iron." Preprint of papers submitted at the 205th ACS National Meeting, Denver, CO.

Matheson, L.J. and P.G. Tratnyek. 1994. "Reductive Dehalogenation of Chlorinated Methanes by Iron Metal." *Environ. Sci. Technol. 28*: 2045-2053.

Pinder, G.F., and L.M. Abriola. 1982. "Calculation of Velocity in Three-Space Dimensions from Hydraulic Head Measurements." *Groundwater 20*(2): 205-209.

SUPPORTING INFORMATION FOR HYDROGEOLOGIC MODELING

B.1 GROUNDWATER FLOW MODEL REVIEW

This appendix presents the general concepts of groundwater flow modeling and describes several modeling codes that may be used in designing and evaluating the permeable barrier systems.

B.1.1 Groundwater Flow Modeling Concepts

To aid in the design of a permeable barrier system and the interpretation of the resulting flow field, it is recommended that a groundwater flow model be constructed using the site-specific geologic and hydrogeologic data collected as part of the site characterization effort. The model can be used to assess the area of influence, optimize the design, and design the performance monitoring network for the permeable barrier system. A complete description of groundwater flow modeling and the mathematics involved is provided in Wang and Anderson (1982) and Anderson and Woessner (1992). The steps involved in model construction and execution are discussed below.

B.1.1.1 Conceptual Model Development. The first step in any modeling effort is the development of the conceptual model. The conceptual model is a three-dimensional (3D) representation of the groundwater flow and transport system based on all available geologic, hydrogeologic, and geochemical data for the site. A complete conceptual model will include geologic and topographic maps of the site, cross sections depicting the site geology/hydrogeology, a description of the physical and chemical parameters associated with the aquifer(s), and contaminant concentration and distribution maps. The purpose of the conceptual model is the integration of the available data into a coherent representation of the flow system to be modeled. The conceptual model is used to aid in model selection, model construction, and interpretation of model results.

B.1.1.2 Model Selection. In order to be used to simulate the flow at permeable barriers, the groundwater flow model requires several special features/ capabilities. The most important requirements derive from the need to simulate sharp hydraulic conductivity (K) contrasts at the intersection of the aquifer and the funnel walls. The specific requirements and recommendations for the permeable barrier simulation models include the following:

- Two-dimensional (2D) or 3D groundwater flow models may be used to simulate the flow system of a site under consideration. A three-dimensional modeling approach is recommended so that the possibility of underflow or overflow and of interactions between the adjacent aquifer can be examined at the permeable barrier and its vicinity. Vertical-flow velocities and travel times will be of critical significance in the design of systems at sites with significant vertical-flow gradients or in cases where the barriers are not keyed into the underlying confining layer.

- The codes should be able to simulate large contrasts in K at the funnel walls. Most of the permeable barrier designs include a reactive cell with K higher than that of the aquifer and flanking funnel walls with extremely low permeability. The funnels may consist of the slurry wall, which can be several feet wide, or the sheet piles, which are usually less than an inch in width. Therefore, at the intersection of the aquifer and the reactive cells, large K contrasts are developed, and many models are unable to solve these problems due to numerical instabilities. In most cases, the funnel walls are simulated by assigning a very low conductivity to the model cells representing the funnel locations. For accurate simulations, the size of these funnel cells should be the same as that of slurry walls. This results in a very small cell size and a large number of cells in the model. The sheet piles are even thinner than the slurry walls and the required cell sizes may be even smaller. To simulate large areas with sufficient resolution near the funnels but larger cells away from the funnels, models capable of incorporating grid blocks of variable size are recommended. Some alternative approaches have been devised to simulate the low-K funnel walls. These are discussed with the appropriate model descriptions in the "Permeable Barrier Simulation Models" section below.

- Many sites have significant heterogeneities, which result in the development of preferential pathways through which most of the groundwater movement occurs. The permeable barrier design itself imparts heterogeneity to the subsurface system. The simulation of these effects requires models that can handle heterogeneity. Most general-purpose analytical models are based on the assumption of homogeneity, but most numerical models can incorporate heterogeneities.

- Many sites have features such as streams, drains, tunnels, or wells in the vicinity of the permeable barrier sites. For example, at some sites, pump-and-treat remediation may be active in the vicinity of the permeable barriers. These situations require the use of models that can simulate the effects of these internal sinks or sources on the permeable barrier systems.

- The results of the model should be amenable to use with the particle-tracking programs so that the capture zones of the permeable barriers can be evaluated. It should also be possible to calculate volumetric flow budgets for the reactive cells.

Many groundwater flow modeling codes currently on the market meet the above requirements. A comprehensive description of nonproprietary and proprietary flow-and-transport modeling codes can be found in the U.S. Environmental Protection Agency document entitled *Compilation of Ground-Water Models* (van der Heijde and Elnawawy, 1993). Depending on the project's needs, the designer of a permeable barrier system may want to apply a contaminant transport code that can utilize the calculated hydraulic-head distribution and flow field from the flow-modeling effort. If flow and transport in the vadose zone are of concern, a coupled or uncoupled, unsaturated/saturated flow and transport model should be considered. The intention of this protocol is not to endorse a specific code, but to suggest a nonproprietary code (that may also be provided privately) that will serve as an example of the type of modeling code that should be used. The proprietary codes are mentioned only if they have been used to simulate the permeable barrier system at a site. The codes that meet most of the requirements for simulation of permeable barrier systems are discussed in Section B.1.2, "Permeable Barrier Simulation Models."

B.1.1.3 Model Construction and Calibration. Model construction consists primarily of converting the conceptual model into the input files for the numerical model. The hydrostratigraphic units defined in the conceptual model can be used to define the physical framework or grid mesh of the numerical model. In both finite-difference (such as MODFLOW) and finite-element models, a model grid is constructed to discretize the lateral and vertical space that the model is to represent. The different hydrostratigraphic units are represented by model layers, each of which is defined by an array of grid cells. Each grid cell is defined by hydraulic parameters (e.g., K, storativity, cell thickness, cell top, bottom) that control the flow of water through the cells.

Model boundaries are simulated by specifying boundary conditions that define the head or flux of water that occurs at the model grid boundaries or edges. Boundary conditions describe the interaction between the system being modeled and its surroundings. Boundary conditions are used to include the effects of the hydrogeologic system outside the area being modeled and also to make possible isolation of the desired model domain from the larger hydrogeologic system. Three types of boundary conditions generally are utilized to describe groundwater flow: specified-head (Dirichlet), specified-flux (Neumann), and head-dependent flux (Cauchy) (Anderson and Woessner, 1992). Internal boundaries or hydrologic stresses, such as wells, rivers, drains, and recharge, may also be simulated using these conditions.

Calibration of a groundwater flow model refers to the demonstration that the model is capable of producing field-measured heads and flows, which are used as the calibration values or targets. Calibration is accomplished by finding a set of hydraulic parameters, boundary conditions, and stresses that can be used in the model to produce simulated heads and fluxes that match field-measured values within a preestablished range of error (Anderson and Woessner, 1992). Model calibration can be evaluated through statistical comparison of field-measured and simulated conditions.

Model calibration often is difficult because values for aquifer parameters and hydrologic stresses typically are known in relatively few locations and their estimates are influenced by uncertainty. The uncertainty in a calibrated model and its input parameters can be evaluated by performing a sensitivity analysis in which the aquifer parameters, stresses, and boundary conditions are varied within an established range. The impact of these changes on the model output (or hydraulic heads) provides a measure of the uncertainty associated with the model parameters, stresses, and boundary conditions used in the model. To ensure a reasonable representation of the natural system, it is important to calibrate with values that are consistent with the field-measured heads and hydraulic parameters. Calibration techniques and the uncertainty involved in model calibration are described in detail in Anderson and Woessner (1992).

B.1.1.4 Model Execution. After a model has been calibrated to observed conditions, it can be used for interpretive or predictive simulations. In a predictive simulation, the parameters determined during calibration are used to predict the response of the flow system to future events, such as the decrease in K over time or the effect of pumping in the vicinity of the permeable barrier. The predictive requirements of the model will determine the need for either a steady-state simulation or a transient simulation, which would accommodate changing conditions and stresses through time. Model output and hydraulic heads can be interpreted through the use of a contouring package and should be applied to particle-tracking simulations to calculate groundwater pathways, travel times, and fluxes through the cell. Establishing travel times through the cell is a key modeling result that can be used to determine the thickness of the permeable cell.

B.1.2 Permeable Barrier Simulation Models

This section describes the various computer simulation codes that meet the minimum requirements for simulations of groundwater flow and particle movement at the permeable barrier sites. Some of the codes already have been used at permeable barrier sites. Nearly all are readily available from the authors or their sponsoring agencies or through resellers. Proprietary codes are included only if they have been applied at a permeable barrier site. Not discussed are advanced programs, such as HST3D (Kipp, 1987), that can simulate the groundwater flow in the vicinity of permeable barriers, but are in fact designed for simulation of more complex processes.

B.1.2.1 MODFLOW and Associated Programs.

Perhaps the most versatile, widely used, and widely accepted groundwater modeling code is that of the U.S. Geological Survey modular, three-dimensional, finite-difference, groundwater flow model, commonly referred to as MODFLOW (McDonald and Harbaugh, 1988). MODFLOW simulates two-dimensional and quasi- or fully three-dimensional, transient groundwater flow in anisotropic, heterogeneous, layered aquifer systems. MODFLOW calculates piezometric head distributions, flowrates, and water balances; it includes modules for flow toward wells, through riverbeds, and into drains. Other modules handle evapotranspiration and recharge. Various textual and graphical pre- and postprocessors available on the market make it easy to use the code and analyze the simulation results. These include GMS (Groundwater Modeling System) (Brigham Young University, 1996), ModelCad (Rumbaugh, 1993), Visual MODFLOW (Waterloo Hydrogeologic, Inc., 1996), and Groundwater Vistas (Environmental Simulations, Inc., 1996).

Additional simulation modules are available through the authors and third parties. One of these is the Horizontal Flow Barrier (HFB) package (Hsieh and Freckleton, 1993). It is especially useful in simulating the funnel-and-gate design. In normal cases, the slurry walls have to be simulated by very small cells of low K, increasing the number of cells in the model dramatically. The HFB package permits the user to assign the sides of certain cells as planes of low K, while still using a larger cell size at the funnel walls. The low-conductivity HFB planes restrict the flow of water into the cells across the faces representing slurry walls or sheet piles. Another useful addition is the ZONEBUDGET package (Harbaugh, 1990), which allows the user to determine the flow budget for any section of the model. This package may be used to evaluate the volumetric flow through the cell for various design scenarios.

The results from MODFLOW can be used in particle-tracking codes, such as MODPATH (Pollock, 1989) and PATH3D (Zheng, 1989), to calculate groundwater paths and travel times. MODPATH is a postprocessing package used to compute 3D groundwater path lines based on the output from steady-state simulations obtained with the MODFLOW modeling code. MODPATH uses a semi-analytical, particle-tracking scheme, based on the assumption that each directional velocity component varies linearly within a grid cell in its own coordinate direction. PATH3D is a general particle-tracking program for calculating groundwater paths and travel times in transient 3D flow fields. The program includes two major segments—a velocity interpolator, which converts hydraulic heads generated by MODFLOW into a velocity field, and a fourth-order Runge-Kutta numerical solver with automatic time-step size adjustment for tracking the movement of fluid particles (van der Heijde and Elnawawy, 1993). A proprietary code, RWLK3D, developed by Battelle (Naymik and Gantos, 1995), also has been used in conjunction with MODFLOW to simulate the particle movement for the pilot-scale permeable cell installed at Moffett Federal Airfield (Battelle, 1996a). This is a 3D transport and particle-tracking code based on the Random Walk approach to solute transport simulation.

B.1.2.2 FLOWPATH. FLOWPATH (Waterloo Hydrogeologic, Inc., 1996) is a 2D, steady-state, groundwater flow and pathline model. The code can simulate confined, unconfined, or leaky aquifers in heterogeneous and anisotropic media. Complex boundary conditions can be simulated. The program output includes simulated hydraulic heads, pathlines, travel times, velocities, and water balances. The funnel walls can be simulated by constructing a model grid with very small cell size in the vicinity of the permeable cells. Because of its user-friendly graphical interface, this program can be used to quickly simulate the flow fields for a number of design options. Therefore, this program has been used for several permeable barrier sites. However, this program cannot be used if the groundwater flow at a site is very complex due to vertical fluxes or if transient flow fields are to be simulated. These situations are possible if there is a potential for vertical underflow or if the permeable wall is not keyed into the confining layer.

B.1.2.3 FRAC3DVS. FRAC3DVS is a 3D, finite-element model for simulating steady-state or transient, saturated or variably saturated, groundwater flow and advective-dispersive solute transport in porous or discretely fractured porous media. The code was developed at the University of Waterloo (Therrien, 1992 and Therrien and Sudicky, 1995) and is being marketed by Waterloo Hydrogeologic, Inc. The code includes preprocessors for grid mesh and input file generation and postprocessors for visualization of the simulation results. This program has many advanced features that are generally not required for simple permeable barrier designs. However, it is included here because the code has been used by Shikaze (1996) to simulate a hypothetical funnel-and-gate design. Further, the solute transport features of this code include the ability to simulate the multispecies transport of straight or branching decay chains. This feature may be used to simulate the reaction progress and daughter product generation in the sequential decay of chlorinated solvents in the permeable cells.

In the work by Shikaze, the impermeable cutoff walls are implemented as 2D planes within the 3D computational domain. This is done by adding "false nodes" wherever impermeable nodes are desired. As a consequence, at the impermeable walls, two nodes exist at the same spatial location. These two nodes are connected to elements on the opposite sides of the wall, essentially breaking the connection between two adjacent elements. The net result is an impermeable wall simulated as a 2D plane within the 3D domain. These simulations assume that the funnel walls are fully impermeable. This may not be a realistic assumption for very long-term simulations, especially for slurry walls.

B.1.2.4 GROWFLOW. GROWFLOW is an innovative permeable barrier simulation program being developed by Applied Research Associates, Inc. (Everhart, 1996) for the U.S. Air Force. The program is based on the Lagrangian smooth particle hydrodynamics (SPH) concepts traditionally used in the astrophysical simulations. SPH is a continuum dynamics solution methodology in which all hydrodynamic and history information is carried on particles. In that sense, GROWFLOW is similar to the particle-tracking codes commonly used to display the flowpaths calculated by the numerical models. The particles in GROWFLOW are Lagrangian interpolation points that interact through the use of a smoothing kernel. The kernel defines a region of influence for each particle and permits approximations to spatial derivatives to be obtained without a mesh. The spatial derivatives are obtained from each particle using an explicit time-integration method.

GROWFLOW is a fully 3D, saturated-unsaturated code that can handle complex geometry. The model domain and the permeable barrier are simulated using exterior and interior flow control panels that contain and direct flow. No model grid is required. Instead, the initial particle locations serve as the integration points for spatial derivatives. The flow control panels form an impermeable boundary that restricts flow across the external model boundaries or across the internal panels that represent funnel walls. The external boundaries are simulated by assigning constant head or constant velocity source models. These source models are panels that control flow into the model domain. The flow out of the model domain is provided by a volume for the fluid to flow into; that is, the model domain is increased.

GROWFLOW input consists of the model domain parameters, the material properties, the elevation head direction, the panel locations, the saturation vs. head relationship, time-step information, saturation vs. conductivity relationship, initial locations of all particles in the system, and particle volume. In addition, information is also needed for the smoothing length (region of influence) for the particles. The output includes a listing of the input parameters, particle locations, and heads at specified time intervals. The output can be plotted to show heads as contour maps and particle movement as pathlines.

GROWFLOW is a highly innovative, flexible, and versatile code for simulation and optimization of permeable barrier systems. However, the code is still under development and several issues need to be addressed. Most importantly, the code needs to be validated against the existing analytical or numerical codes and against field data to verify its numerical accuracy. There appears to be no clear method for simulating internal sources or sinks such as wells and rivers. At many sites, these features may form a significant part of the hydrologic budgets. In addition, there appears to be no provision to check mass or volume balance in the simulations.

B.1.2.5 Funnel-and-Gate Design Model (FGDM). FGDM is a multicomponent, steady-state, analytical program for funnel-and-gate design and cost optimization. It was developed by Applied Research Associates (Hatfield, 1996) for the U.S. Air Force. Program input includes the initial concentrations and first-order reaction rates and the required water quality standards. These are used to determine the required residence times for water in the permeable cell. The critical residence times are used by the program along with input-plume-to-gate-width ratios to develop several funnel-and-gate designs. Finally, the cost minimization model is used to find the minimum cost design scenario based on the input unit costs for funnel walls, gate walls, reactive media, and land. The Lagrangian cost minimization is based on a modified Newton-Raphson algorithm for solution of nonlinear equations. Because the accuracy of cost minimization is based partly on the initial estimates for the minimum cost design, it is important to have a preliminary estimate of the low-cost configuration. Additional input parameters include the funnel width, hydraulic gradient, aquifer thickness, $K_{aquifer}$, gate porosity, ratio of aquifer to K_{cell}, and depth of system walls. The funnel width, the total width of funnel walls and the gate, is estimated in advance assuming a capture efficiency of 80%. For example, for a plume width of 80 ft, a funnel width of 100 ft is suggested. This assumption may need to be validated by further modeling or field studies. FGDM is a useful tool for a quick evaluation of several design scenarios in a simple setting. However, it cannot be used for complex settings such as heterogeneous media, or for evaluating the flowpaths through the permeable cell.

B.1.2.6 FLONET. FLONET (Guiguer et al., 1992) is a 2D, steady-state flow model distributed by Waterloo Hydrogeologic, Inc. The program calculates potentials, streamlines, and velocities and can be used to generate flownets (maps showing flowlines and hydraulic heads) for heterogeneous, anisotropic aquifers. The funnel walls and the gate can be specified by assigning lower K to elements representing these features. The program was used by Starr and Cherry (1994) to evaluate several design scenarios for funnel-and-gate systems.

B.1.3 Previous Modeling Studies for Permeable Barrier Applications

A review of the information available from prevailing sites showed that MODFLOW (McDonald and Harbaugh, 1988) in conjunction with particle tracking with codes such as MODPATH (Pollock, 1989), is the code most commonly used to simulate the permeable barriers technology. Other programs such as FLONET (Guiguer et al., 1992), FRAC3DVS (Therrien and Sudicky, 1995), FLOWPATH (Waterloo Hydrogeologic, Inc., 1996), and RWLK3D (Naymik and Gantos, 1995) also have been used at some sites. Two new codes, GROWFLOW (Everhart, 1996) and FGDM (Funnel and Gate Design Model) (Hatfield, 1996) have been developed recently for the U.S. Air Force to simulate and optimize the funnel-and-gate systems. However, these new codes have so far not been applied at any sites. The sites that used MODFLOW include the Sunnyvale, California site, Moffett Federal Airfield, California (PRC, 1996 and Battelle, 1996a), the Sommersworth Sanitary Landfill, New Hampshire, an industrial facility in Kansas, and GE Appliances, Wisconsin. FLOWPATH has been used to evaluate the design at Belfast, Northern Ireland; Fairchild Air Force Base, Washington; and the DOE Kansas City site, Kansas. The names of simulation codes used at other sites were not readily available. The most comprehensive modeling evaluations of the permeable barrier technology are those by Starr and Cherry (1994), and Shikaze (1996). These papers evaluate the effects of various parameters on the design and performance of hypothetical funnel-and-gate configurations, although some of the conclusions are applicable to continuous reactive barriers as well.

Starr and Cherry (1994) used a 2D, plan-view, steady-state flow simulation program, FLONET (Guiguer et al., 1992) to illustrate the effects of funnel-and-gate geometry (design) and reactive cell hydraulic conductivity (K_{cell}) on the size and shape of capture zone, the discharge groundwater flow volume through the gate, and the residence time in the reactive cell. Only the configurations with barriers that penetrate the entire aquifer thickness and extend into the underlying confining layer were simulated. The hanging wall systems were not simulated because they can best be described by 3D simulations. The simulated system had properties similar to those of the surficial aquifer at Canadian Forces Base Borden, Ontario, Canada. The simulated aquifer is isotropic, with homogeneous aquifer hydraulic conductivity ($K_{aquifer}$) of 28.3 feet/day and hydraulic gradient of 0.005. The funnel walls were assumed to be 1-m (3.28-feet)-thick slurry walls with K equal to 0.0028 feet/day. The K of the reactive cell was 283 feet/day, the maximum laboratory-measured value for 100 percent iron, in the base case. It should be noted that in several other modeling studies for permeable cell installations, K_{cell} values of 142 feet/day have been used for 100 percent iron. The range of values for K_{cell} indicates differences in the source of granular iron, as well as variability of the K measurement itself. A porosity of 0.33 was used for all materials. The following conclusions were made by these researchers based on the simulation of several scenarios.

- For systems with funnel walls at 180 degrees (straight funnel), the discharge through the gate and the hydraulic capture zone width increases as the funnel width increases. However, the increase in discharge is not directly proportional to funnel width. In fact, the relative discharge through the gate decreases dramatically as the funnel width increases. Relative discharge refers to the ratio of discharge through the gate to the discharge through the area in the absence of the funnel-and-gate system.

- For a constant funnel width, the absolute and relative discharge through the gate (and the capture zone width) increase with an increase in gate width. Therefore, it is desirable to have a gate as wide as practical.

- For a given funnel-and-gate design, the discharge through the gate increases with increase in K_{cell} relative to the $K_{aquifer}$. However, there is relatively little increase in discharge when the K_{cell} is more than 10 times higher than the $K_{aquifer}$. This implies that while a reactive cell conductivity higher than the $K_{aquifer}$ is desirable, K_{cell} does not have to be much higher than $K_{aquifer}$. This is a useful result, because the large grain sizes required for very high-K_{cell} values would result in a low total surface area for reactions and lower residence times.

- For all orientations to the regional flow gradient, the maximum absolute discharge occurs at apex angles (the angles between the two funnel walls) of 180 degrees (straight barrier). However, for apex angles between 127 and 233 degrees there is little effect on discharge. Outside this range, the discharge drops rapidly. This implies that there is no significant advantage of a slightly angled funnel-and-gate system over a straight barrier and vice versa. Sharper funnel angles may, however, reduce discharge.

- For all apex angles, the maximum discharge occurs when the funnel is perpendicular to the regional flow gradient.

- The groundwater flow models can be used effectively to design the funnel-and-gate systems at sites with special design requirements due to complex flow fields, seasonal fluctuations, or access restrictions. These may include systems with angled funnels, multiple gates, asymmetrical funnels, or U-shaped funnel and gates.

- A balance between maximizing the capture zone of the gate and maximizing the residence times of contaminated water in the gate should be achieved. In general, the discharge and residence times are inversely proportional. The residence time can generally be increased without affecting the capture zone by increasing the width of the gate.

Shikaze (1996) used FRAC3DVS code to examine 3D groundwater flow in the vicinity of a partially penetrating (hanging wall) funnel-and-gate system for 16 different combinations of parameters. All simulations were for steady-state, fully saturated groundwater flow. The 16 simulations consisted of variations in four dimensionless parameters: the ratio of K_{cell} to $K_{aquifer}$; the ratio of width of a single funnel wall to the depth of the funnel-and-gate; the ratio of total funnel wall width to the gate width; and the hydraulic gradient. The following conclusions were drawn from these simulations:

- Absolute discharge through the gate increases as the hydraulic gradient increases. However, there is almost no effect of hydraulic gradient on the relative discharge or on the size of the relative capture zone (hydraulic capture zone width ÷ total width of funnel-and-gate).

- For higher values of K_{cell} versus $K_{aquifer}$, there is an increase in absolute and relative discharge through the gate as well as in the relative size of the capture zone. Thus, a higher K_{cell} tends to draw more flow towards the gate.

- Higher values for the ratio of width of the single funnel wall (one wing) to the depth of the funnel-and-gate system result in lower absolute and relative discharge, and in smaller capture zones. This is due to the fact that in cases of wide but shallow funnel walls, there is an increase in the flow component that is diverted under the barrier rather than through the gate.

- Higher values for the ratio of total funnel wall width to the width of the gate result in higher absolute discharge but lower relative discharge and smaller hydraulic capture zones. This implies that, for wider funnel walls, the increase in the discharge through the gate is not proportional to the increase in the funnel wall area.

B.2 ILLUSTRATION OF THE HYDROLOGIC MODELING APPROACH FOR PERMEABLE BARRIER APPLICATION

The following methodology serves as an illustration of the permeable barrier design modeling approach for homogeneous and heterogeneous aquifers. Modeling may be used to design the location, configuration, and dimensions of the permeable barrier, as well as to develop a performance monitoring plan.

B.2.1 Homogeneous Aquifers

MODFLOW can be used to develop a steady-state numerical approximation of the groundwater flow field and to calculate flow budgets through the gate. Particle-tracking techniques under advective flow conditions only can be used to delineate capture zones and travel times in the vicinity of the funnel and gate. RWLK3D (Prickett et al., 1981) or any similar particle tracking code could be used to simulate particle pathways. The model simulations can be performed to aid in both the design phase and the evaluation phase of permeable barrier systems for the containment and remediation of contaminated groundwater. These simulations can build upon previous modeling efforts conducted by Starr and Cherry (1994). Specific objectives can include determining how changes in gate conductivity over time affected capture zone width, retention times for groundwater moving through the gate, and flow volumes through the gate.

The model domain and grid size typically are determined based on the site-specific conditions. The primary criteria are that the domain should be large enough so that the boundary conditions do not affect flow in the vicinity of the permeable barrier. Further, the model cell size in the vicinity of the permeable barrier should be small enough to provide sufficient resolution for retention time calculations. The funnel-and-gate configuration modeled in this illustration is a pilot barrier at a U.S. Navy base in California (see Figure B-1). The funnel consists of two 20-foot lengths of sheet piling oriented perpendicular to flow on either side of a 10-foot by 10-foot reactive cell representing the gate. The reactive cell is bounded on its sides by 10-foot lengths of sheet piling. The gate itself consists of 2 feet of ¾-inch pea gravel located on both the upgradient and downgradient ends of the reactive cell, which has a 6-foot flowthrough thickness.

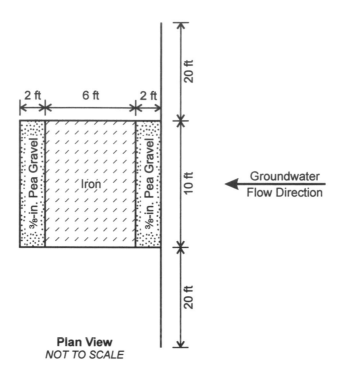

Plan View
NOT TO SCALE

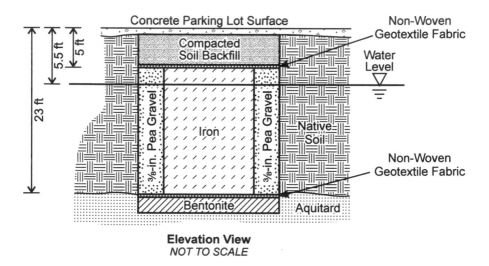

Elevation View
NOT TO SCALE

FIGURE B-1. Pilot-scale funnel-and-gate system installed at Moffett Federal Airfield, CA

For this model of a funnel-and-gate system, the domain consisted of a single layer that is 500 feet long and 300 feet wide. The grid has 98 rows and 106 columns resulting in a total of 10,388 nodes. Grid nodes are 10 feet by 10 feet at their maximum (in the general domain area) and 0.5 foot by 0.5 foot in the region of the gate itself. Specified head nodes were set along the first and last rows of the model to establish a gradient of 0.006. No flow conditions were set along the first and last columns of the model.

The funnel (sheet piling) was simulated as a horizontal flow barrier having a hydraulic conductivity (K) of 2.0×10^{-6} feet/day. For the continuous reactive barrier configuration, the funnel may be excluded from the model. The pea gravel was assigned a K of 2,830 ft/d. The reactive cell consisting of granular iron was assigned a K of 283 ft/d, the maximum laboratory-measured value for 100% iron. It should be noted that in some modeling studies (e.g., Thomas et al., 1995), a reactive cell with K of 142 ft/d has been used for 100% iron. In general, the K value for the reactive medium should be determined from laboratory permeability testing. Porosity was held constant at 0.30 for all materials in each of the simulations.

For this illustration, simulated $K_{aquifer}$ was varied among 0.5, 1, 2, 5, 10, 20, 50, and 100 ft/d to represent low- and high-permeability aquifers. Once this base scenario was established, simulations were conducted to evaluate reductions in K_{cell} over time that could potentially be caused by buildup of precipitates. To determine the effects of decreased permeability of the gate over a period of operation, K_{cell} was reduced in 10 percent increments from the initial 283 ft/d to 28.3 ft/d for each value of $K_{aquifer}$. An additional set of simulations was performed with K_{cell} reduced by 95% to 14.15 ft/d, resulting in a total of 11 simulations for each value of $K_{aquifer}$. For each individual simulation, a single value for $K_{aquifer}$ was used. The effects of geologic heterogeneities were not considered in these simulations. The results from the 88 simulations were used to evaluate the impact of variations in K_{cell} and $K_{aquifer}$ on capture zone width, flow volumes, and travel times (retention time) through the gate.

Table B-1 lists the model run number, gate conductivity, aquifer conductivity, ratio of reactive cell to aquifer conductivity, capture zone width, residence time within the reactive cell, and groundwater discharge through the gate. Capture zone width in each of the simulations was determined by tracking particles forward through the gate. Two hundred particles (1 particle every 0.5 feet) were initiated along a 100-foot-long line source upgradient from the barrier. The locations of the flow divides between particles passing through the gate and those passing around the ends of the funnel were used to determine capture zone width. Residence time within the gate for each simulation was determined from the length of time required for the particles to pass through the reactive cell. Figure B-2 illustrates the determination of flow divides and travel times for simulation number 57, which had an aquifer conductivity of 20 ft/d and a reactive cell conductivity of 283 ft/d. Particle pathlines have been overlain upon the calculated water-table surface. Particle pathlines and intermediate time steps within the reactive cell are also shown. In some cases, there may be significant variation in residence times at

TABLE B-1. Summary of funnel-and-gate model runs

Run #	K_{gate} (ft/day)	$K_{aquifer}$ (ft/day)	$K_g:K_{aq}$	Capture Width (ft)	Discharge Cu ft/day	Residence Time (days)	Relative Discharge
1	283	0.1	2830.00	NA	NA	NA	NA
2	283	0.5	566.00	34	2.356	219.0	1.000
3	255	0.5	509.40	NA	2.356	220.0	1.000
4	226	0.5	452.80	NA	2.355	218.0	1.000
5	198	0.5	396.20	NA	2.355	219.0	1.000
6	170	0.5	339.60	NA	2.354	220.0	0.999
7	142	0.5	283.00	NA	2.354	219.0	0.999
8	113	0.5	226.40	NA	2.353	218.0	0.999
9	85	0.5	169.80	NA	2.352	220.0	0.998
10	57	0.5	113.20	NA	2.350	220.0	0.998
11	28	0.5	56.60	NA	2.344	220.0	0.995
12	14	0.5	28.30	NA	2.334	NA	0.991
13	283	1	283.00	32.75	4.732	107.0	1.000
14	255	1	254.70	NA	4.732	107.5	1.000
15	226	1	226.40	NA	4.730	107.5	1.000
16	198	1	198.10	NA	4.729	107.5	0.999
17	170	1	169.80	NA	4.727	107.5	0.999
18	142	1	141.50	NA	4.725	107.5	0.998
19	113	1	113.20	NA	4.721	107.5	0.998
20	85	1	84.90	NA	4.716	107.5	0.997
21	57	1	56.60	NA	4.705	108.0	0.994
22	28	1	28.30	NA	4.672	108.5	0.987
23	14	1	14.15	NA	4.603	110.0	0.973
24	283	2	141.50	NA	9.475	52.5	1.000
25	255	2	127.35	NA	9.472	52.5	1.000
26	226	2	113.20	NA	9.468	52.5	0.999
27	198	2	99.05	NA	9.462	52.5	0.999
28	170	2	84.90	NA	9.455	52.5	0.998
29	142	2	70.75	NA	9.446	52.5	0.997
30	113	2	56.60	NA	9.432	53.0	0.995
31	85	2	42.45	NA	9.408	53.0	0.993
32	57	2	28.30	NA	9.362	53.5	0.988
33	28	2	14.15	NA	9.223	54.5	0.973
34	14	2	7.08	NA	8.954	56.0	0.945
35	283	5	56.60	32.17	23.613	21.0	1.000
36	255	5	50.94	NA	23.593	20.9	0.999
37	226	5	45.28	NA	23.568	21.0	0.998
38	198	5	39.62	NA	23.535	21.1	0.997
39	170	5	33.96	NA	23.493	21.1	0.995
40	142	5	28.30	NA	23.432	21.1	0.992
41	113	5	22.64	NA	23.344	21.3	0.989
42	85	5	16.98	NA	23.197	21.4	0.982
43	57	5	11.32	NA	22.909	21.6	0.970
44	28	5	5.66	NA	22.082	22.6	0.935
45	14	5	2.83	NA	20.597	24.0	0.872
46	283	10	28.30	32.17	46.407	10.6	1.000
47	255	10	25.47	32.17	46.328	10.6	0.998
48	226	10	22.64	32.17	46.169	10.8	0.995
49	198	10	19.81	32.33	46.040	10.7	0.992
50	170	10	16.98	32.33	45.870	10.9	0.988
51	142	10	14.15	32.5	45.628	10.9	0.983
52	113	10	11.32	31.5	45.274	11.0	0.976
53	85	10	8.49	31.67	44.763	11.0	0.965
54	57	10	5.66	31.83	43.566	11.4	0.939
55	28	10	2.83	32.17	40.562	12.3	0.874
56	14	10	1.42	NA	35.630	13.9	0.768
57	283	20	14.15	31.81	91.493	5.4	1.000
58	255	20	12.74	NA	91.239	5.4	0.997
59	226	20	11.32	NA	91.331	5.5	0.998
60	198	20	9.91	NA	89.890	5.6	0.982

TABLE B-1. Summary of funnel-and-gate model runs (cont'd)

Run #	K_{gate} (ft/day)	$K_{aquifer}$ (ft/day)	$K_g:K_{aq}$	Capture Width (ft)	Discharge Cu ft/day	Residence Time (days)	Relative Discharge
61	170	20	8.49	NA	89.262	5.6	0.976
62	142	20	7.08	NA	88.379	5.6	0.966
63	113	20	5.66	NA	86.708	5.7	0.948
64	85	20	4.25	NA	84.126	5.8	0.919
65	57	20	2.83	NA	78.681	6.3	0.860
66	28	20	1.42	NA	73.403	6.7	0.802
67	14	20	0.71	NA	59.502	8.3	0.650
68	283	50	5.66	31.5	221.445	2.3	1.000
69	255	50	5.09	NA	219.770	2.3	0.992
70	226	50	4.53	NA	217.730	2.3	0.983
71	198	50	3.96	NA	215.185	2.4	0.972
72	170	50	3.40	NA	211.925	2.4	0.957
73	142	50	2.83	NA	207.005	2.4	0.935
74	113	50	2.26	NA	200.755	2.5	0.907
75	85	50	1.70	NA	190.560	2.6	0.861
76	57	50	1.13	NA	173.695	2.9	0.784
77	28	50	0.57	NA	136.155	3.7	0.615
78	14	50	0.28	NA	94.409	5.8	0.426
79	283	100	2.83	30.38	410.105	1.3	1.000
80	255	100	2.55	NA	404.240	1.2	0.986
81	226	100	2.26	NA	397.135	1.2	0.968
82	198	100	1.98	NA	388.355	1.3	0.947
83	170	100	1.70	NA	377.240	1.3	0.920
84	142	100	1.42	NA	362.735	1.4	0.884
85	113	100	1.13	NA	343.060	1.5	0.837
86	85	100	0.85	NA	314.455	1.6	0.767
87	57	100	0.57	NA	268.935	1.8	0.656
88	28	100	0.28	NA	188.075	2.7	0.459
89	14	100	0.14	NA	116.935	4.2	0.285
90	283	200	1.42	NA	NA	NA	NA

the edges of the reactive cell and at its center. For example, Vogan et al. (1994) showed that simulated residence times in a funnel-and-gate system (with caisson gates) varied from 29 hours at the edges to 82 hours in the center of the reactive cell.

Discharge through the gate was determined from the MODFLOW-calculated, cell-by-cell flow file using the MODUTILITY code zone budget (Harbaugh, 1990). Correlations between $K_{aquifer}$ and K_{cell}, retention time, discharge, and capture zone width can be determined by plotting the results of the 88 simulations against one another. Some basic relationships are readily apparent.

Figure B-3 illustrates the correlation between $K_{aquifer}$, retention time, and discharge through the gate. There is an inverse relationship between $K_{aquifer}$ and retention time. As aquifer conductivity increases, the retention time within the gate decreases. As aquifer conductivity increases, the total discharge through the gate increases. Finally, Figure B-3 shows a very strong inverse correlation between the total discharge through the gate and the retention time within the gate. Therefore, aquifers having high hydraulic conductivities may require greater flowthrough thickness of gate to meet residence time requirements so that contaminant levels can be reduced to regulatory limits.

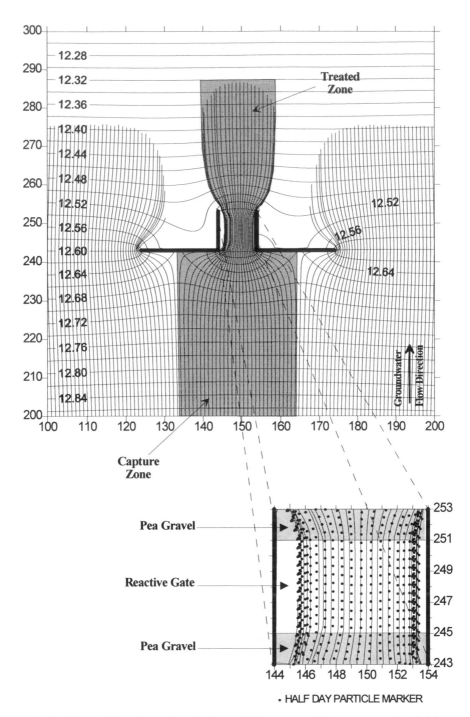

• HALF DAY PARTICLE MARKER

FIGURE B-2. Simulated particle pathlines overlain upon water table including zoomed-in view of gate area

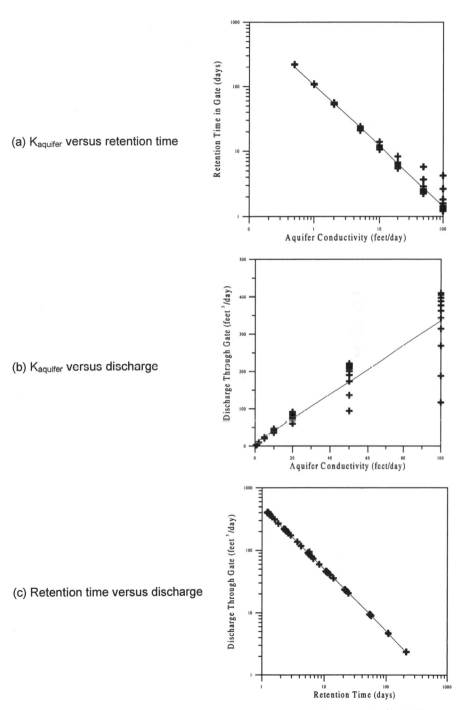

(a) $K_{aquifer}$ versus retention time

(b) $K_{aquifer}$ versus discharge

(c) Retention time versus discharge

FIGURE B-3. Correlation between $K_{aquifer}$, discharge, and travel time through the gate for a homogeneous, one-layer scenario

The conductivities of both the aquifer and the reactive cell were plotted against capture zone width. A general correlation exists between an increase in K (and discharge through the gate) and capture zone width. As K increased, the capture zone width generally increased. However, the capture zone width appears to be more sensitive to the length of the funnel walls and was generally observed to occur at just over half of the funnel wall length on either side of the gate. Capture zone widths ranged from roughly 0.2 to 2 feet beyond the midpoint of the funnel wall. Figure B-4 is a plot showing the reduction in discharge (due to potential buildup of precipitate) through the gate that results from decreasing K_{cell} at aquifer conductivities of 0.5, 10, and 100 ft/d. In each of the plots shown in Figure B-4, K_{cell} decreases from 283 ft/d to 14.15 ft/d. Reductions in K_{cell} were simulated to represent the potential clogging of the reactive cell by precipitation. The percent decline in discharge through the gate was determined for each decline

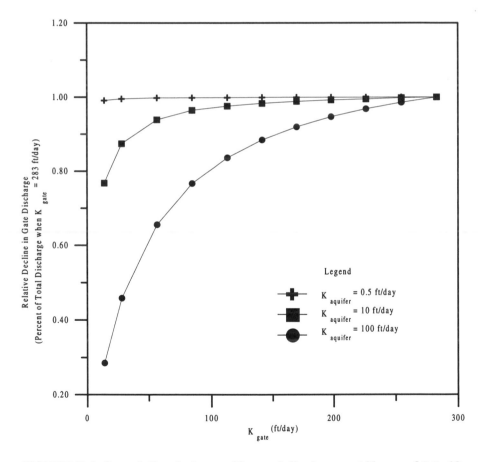

FIGURE B-4. Correlation between K_{cell} and discharge at $K_{aquifer}$ of 0.5, 10, and 100 feet per day K_{cell} varied between 283 and 14.15 feet per day

in K_{cell}. When aquifer conductivity is 0.5 ft/d, the reactive cell conductivity is much greater than the aquifer conductivity for each of the 11 simulations performed, and the percent decline in discharge through the gate is very small. Decreasing reactive cell conductivity from 283 ft/d to 14.15 ft/d resulted in only a 1 percent decline in the discharge through the gate. As aquifer conductivity was increased, a larger reduction in discharge through the gate occurred as the reactive cell conductivity decreased. For aquifer conductivities of 10 and 100 ft/d, discharge through the gate decreased by roughly 27 and 71 percent, respectively, over the same decline in gate conductivity. In both cases, the ratio of K_{cell} to $K_{aquifer}$ approaches or becomes less than 1 as K_{cell} decreases. Therefore, the effects of precipitate buildup in the reactive cell are likely to be felt earlier in high-permeability aquifers. However, as discussed below, there is considerable leeway before such effects are noticed.

Figure B-5 is a plot of the ratio of K_{cell} to $K_{aquifer}$ versus discharge through the gate for the 88 simulations. The plot indicates that declines in reactive cell

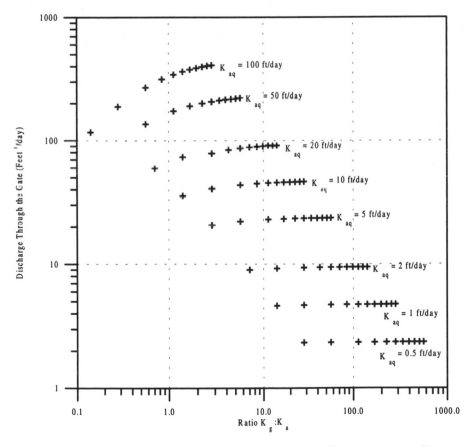

FIGURE B-5. Correlation between ratio of K_{cell} to $K_{aquifer}$ versus discharge through the gate for a homogeneous, one-layer scenario

conductivity due to clogging have very little influence on the volume of ground-water passing through the gate as long as the reactive cell conductivity is roughly 5 times the conductivity of the aquifer. In these instances, discharge through the gate remained at roughly 95 percent of the simulated discharge when the gate conductivity was 283 ft/d. Because discharge is relatively unaffected, residence times and capture zone width will remain relatively unchanged for a given aquifer conductivity. As the ratio between K_{cell} and $K_{aquifer}$ declines below 5, the relative decrease in discharge becomes greater and results in decreased capture zone widths and increased retention times. Thus, as long as the hydraulic conductivity of a freshly installed reactive cell is designed to be one or two orders of magnitude greater than the hydraulic conductivity of the aquifer, there is considerable flexibility for precipitates to build up without significantly affecting the hydraulic capture zone.

B.2.2 Heterogeneous Aquifers

Most modeling studies at previous permeable barrier sites were based on the assumption that the aquifer sediments in the vicinity of the permeable barrier are homogeneous. However, at many sites, there may be strong heterogeneity in the sediments. This heterogeneity develops mainly due to the variations in deposi-tional environments of the sediments. The general implications of heterogeneity are that more detailed site characterization is required and the models are more complex. The symmetrical capture zones seen in cases of homogeneous sediments become asymmetrical and difficult to predict without detailed characterization and modeling.

Examples of the effect of heterogeneity on the flowpaths and capture zones can be seen from the modeling work conducted in support of the design and performance monitoring for the Moffett Federal Airfield (MFA) Site (Battelle, 1996b and PRC, 1996) and the Elizabeth City, North Carolina site (Puls et al., 1995). Groundwater flow modeling for the MFA pilot barrier showed that the presence of heterogeneities due to multiple subsurface channels (strata) causes the capture zones to be substantially asymmetrical. Figure B-6 is a simulated flowpath diagram showing the result of backward particle tracking for 25 days with parti-cles starting from the funnel area in model layers 1 through 4 at the funnel location. The reactive cell is present in layers 2, 3, and 4 of the model.

The most striking observation from this figure is that the capture zone for a permeable barrier at a heterogeneous site is highly asymmetrical and there is a significant difference in the residence time at different depth levels. For example, there is almost no movement of particles in 25 days in layers 1 and 2. In layer 3, the particle movement is very fast directly upgradient of the gate but very slow upgradient of the funnel walls. In layer 4, the particle movement is very fast upgradient of the gate in the west funnel wall but still very slow upgradient of the east funnel. These differences in particle velocities and resulting irregularities in the capture zones are because the lower part of the reactive cell is located in a high-permeability sand channel, whereas the funnel walls and the upper portion of the reactive cell are located in the lower conductivity interchannel deposits. The

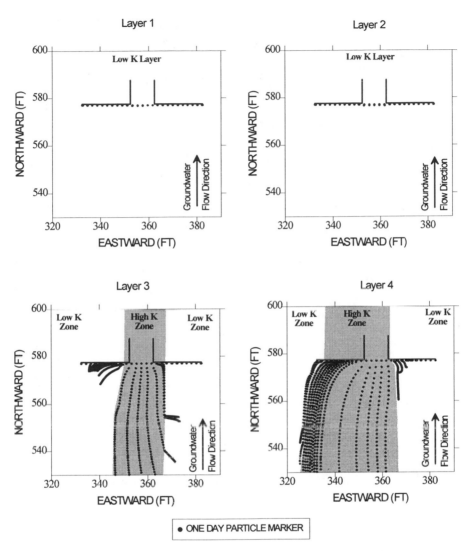

FIGURE B-6. MFA funnel-and-gate backward particle tracking showing the effect of heterogeneity on capture zones (Battelle, 1996b)

location of sand channels at the site was determined based on the preexisting base-wide site characterization maps and from site-specific CPT data.

At the Elizabeth City, North Carolina site (Puls et al., 1995), the geology is characterized by complex and variable sequences of surficial sands, silts, and clays. Groundwater flow velocity is extremely variable with depth, with a highly conductive layer at approximately 12 to 20 feet below ground surface. The reactive metal zone was emplaced in this sand channel (Figure B-7).

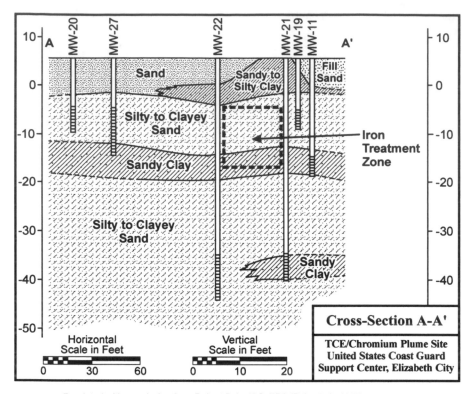

Reprinted with permission from Robert Puls, U.S. EPA (Puls et al., 1995).

FIGURE B-7. Location of reactive cell in a sand channel

These examples illustrate the need for placing the reactive cell in a zone of high conductivity that forms a preferential pathway for most of the flow and contaminant transport through the aquifer. Additionally, the dependence of capture zones on aquifer heterogeneities illustrates the need for detailed site characterization prior to permeable barrier placement.

B.3 APPENDIX B REFERENCES

Anderson, M.P. and W.W. Woessner. 1992. *Applied Groundwater Modeling: Simulation of Flow and Advective Transport.* Academic Press, New York, NY.

Battelle. 1996a. Evaluation of Funnel-and-Gate Pilot Study at Moffett Federal Airfield with Groundwater Modeling. Draft, prepared for the U.S. Department of Defense, Environmental Security Technology Certification Program and Naval Facilities Engineering Service Center, Port Hueneme, CA. September 11.

Battelle. 1996b. Performance Monitoring Plan, A Reactive Wall at Moffett Federal Airfield. Draft, prepared for the U.S. Department of Defense, Environmental

Security Technology Certification Program and Naval Facilities Engineering Service Center, Port Hueneme, CA. September 16.

Brigham Young University. 1996. GMS: Department of Defense Groundwater Modeling System, Version 2.0.

Environmental Simulations, Inc. 1996. Groundwater Vistas modeling software.

Everhart, D. 1996. *Theoretical Foundations of GROWFLOW.* ARA-TR-96-5286-3. Prepared by Applied Research Associates, Inc. for U.S. Air Force, Tyndall Air Force Base, FL, April.

Guiguer, N., J. Molson, E.O. Frind, and T. Franz. 1992. *FLONET—Equipotential and Streamlines Simulation Package.* Waterloo Hydrogeologic Software and the Waterloo Center for Groundwater Research, Waterloo, Ontario.

Harbaugh, A.W. 1990. *A Computer Program for Calculating Subregional Water Budgets Using Results from the U.S. Geological Survey Modular Three-Dimensional Finite-Difference Ground-Water Flow Model.* U.S. Geological Survey Open-File Report 90-392.

Hatfield, K. 1996. *Funnel-and-Gate Design Model.* ARA-TR-96-5286-4. Prepared by Applied Research Associates, Inc. for U.S. Air Force, Tyndall Air Force Base, FL, April.

Hsieh, P.A., and J.R. Freckleton. 1993. *Documentation of a Computer Program to Simulate Horizontal-Flow Barriers Using the U.S. Geological Survey Modular Three-Dimensional Finite-Difference Ground-Water Flow Model.* U.S. Geological Survey Open-File Report 92-477.

Kipp, Jr., K.L. 1987. *HST3D. A Computer Code for Simulation of Heat and Solute Transport in Three-Dimensional Groundwater Flow Systems.* WRI 86-4095. U.S. Geological Survey, Denver, CO.

McDonald, M.G., and A.W. Harbaugh. 1988. *A Modular Three-Dimensional Finite-Difference Ground-Water Flow Model: Techniques of Water-Resources Investigations of the United States Geological Survey.* Book 6.

Naymik, T.G., and N.J. Gantos. 1995. Solute Transport Code Verification Report for RWLK3D, Internal Draft. Battelle Memorial Institute, Columbus, OH.

Pollock, D.W. 1989. *Documentation of Computer Programs to Compute and Display Pathlines Using Results from the U.S. Geological Survey Modular Three-Dimensional Finite-Difference Ground-Water Flow Model.* U.S. Geological Survey Open-File Report 89-381.

PRC Environmental Management, Inc. 1996. *Naval Air Station Moffett Field, California, Iron Curtain Area Groundwater Flow Model.* PRC, June.

Prickett, T.A., T.G. Naymik, and C.G. Lounquist. 1981. *A "Random Walk" Solute Transport Model for Selected Groundwater Quality Evaluations.* Illinois Department of Energy and Natural Resources, Illinois State Water Survey, Bulletin 65.

Puls, R.W., R.M. Powell, and C.J. Paul. 1995. "In Situ Remediation of Ground Water Contaminated with Chromate and Chlorinated Solvents Using Zero-Valent Iron: A Field Study." *Extended Abstracts from the 209th ACS National Meeting Anaheim, CA, 35*(1): 788-791. Anaheim, CA. Division of Environmental Chemistry, American Chemical Society, Washington, DC.

Rumbaugh, J.O., III. 1993. ModelCad[386]: Computer-Aided Design Software for Ground-Water Modeling, ver. 2.0, Geraghty & Miller, Inc., Reston, VA.

Shikaze, S. 1996. *3D Numerical Modeling of Groundwater Flow in the Vicinity of Funnel-and-Gate Systems.* ARA-TR-96-5286-1. Prepared by Applied Research Associates, Inc. for U.S. Air Force, Tyndall Air Force Base, FL, April.

Starr, R.C., and J.A. Cherry. 1994. "In Situ Remediation of Contaminated Ground Water: The Funnel-and-Gate System." *Groundwater 32*(3): 465-476.

Therrien, R. 1992. Three-Dimensional Analysis of Variably-Saturated Flow and Solute Transport in Discretely-Fractured Porous Media. Ph.D. thesis, Dept. of Earth Science, University of Waterloo, Ontario, Canada.

Therrien, R., and E. Sudicky. 1995. "Three-Dimensional Analysis of Variably-Saturated Flow and Solute Transport in Discretely-Fractured Porous Media." *Journal of Contaminant Hydrology 23*: 1-44.

Thomas, A.O., D.M. Drury, G. Norris, S.F. O'Hannesin, and J.L. Vogan. 1995. "The In-Situ Treatment of Trichloroethene-Contaminated Groundwater Using a Reactive Barrier—Result of Laboratory Feasibility Studies and Preliminary Design Considerations." In Brink, Bosman, and Arendt (Eds.), *Contaminated Soil '95*, pp. 1083-1091. Kluwer Academic Publishers.

van der Heijde, P.K.M., and O.A. Elnawawy. 1993. *Compilation of Groundwater Models.* EPA/600/2-93/118, U.S. EPA, R.S. Kerr Environmental Research Laboratory, Ada, OK.

Vogan, J.L., J.K. Seaberg, B.G. Gnabasik, and S. O'Hannesin. 1994. Evaluation of In-Situ Groundwater Remediation by Metal Enhanced Reductive Dehalogenation Laboratory Column Studies and Groundwater Flow Modeling. Unpublished.

Wang, H.F., and M.P. Anderson. 1982. *Introduction to Groundwater Modeling: Finite Difference and Finite Element Methods.* W.H. Freeman and Company, New York, NY.

Waterloo Hydrogeologic, Inc. 1996. *FLOWPATH Users Manual, Version 5.2.* Waterloo, Ontario.

Zheng, C. 1989. *PATH3D.* S. S. Papadopulos and Assoc., Rockville, MD.

SUPPORTING INFORMATION FOR GEOCHEMICAL MODELING

C.1 GEOCHEMICAL MODELS APPLICABLE TO PERMEABLE BARRIERS

Geochemical modeling is an attempt to interpret or predict the concentrations of dissolved species in groundwater based on assumed chemical reactions. Early efforts were concerned with performing speciation calculations on dissolved inorganic constituents. The models that were developed as an outgrowth of these efforts can be grouped as forward and inverse models. In *forward modeling*, reaction progress is governed by thermodynamic expressions; hence, the result is an equilibrium prediction. In *inverse modeling*, probable reactions are calculated based on the information supplied at initial and final points along a flowpath, and as such, do not necessarily represent equilibrium.

The third type, *reaction-transport modeling*, couples forward modeling with fluid flow and solute transport. This is the newest area of geochemical modeling research, and few highly sophisticated codes have been developed in this area. Most reaction-transport models offer less than three-dimensional flow fields, have limited capabilities for introducing heterogeneities in the flow regime, and tend to consider only static boundary conditions. A recent review of geochemical modeling codes can be found in Mangold and Tsang (1991). In addition, up-to-date information on progress in this area can be obtained by contacting the International Ground Water Modeling Center in Golden, Colorado.

Computer codes typically are used to perform numerical algorithms that model chemical reactivity, hydrochemical transport, and in some cases both. Table C-1 provides names and other information for several geochemical modeling codes that are in common usage. Codes that model chemical reactivity typically include speciation of dissolved constituents, redox calculations, adsorption, precipitation and dissolution of minerals or compounds, and mass transfer through exchange with the atmosphere (open vs. closed system).

Factors associated with choosing forward, inverse, and reaction-transport modeling depend on the nature of the geochemical system being considered. Forward modeling may be preferred when only the final outcome of the interaction of groundwater with soils or sediments is desired, i.e., groundwater composition and mineral saturation index. Inverse modeling provides hydrologic information about an aquifer, such as net mass transfer and mixing, and can be used to determine relative rates of reactions. Most models allow testing only to see if expected reactions

TABLE C-1. Geochemical modeling codes

Name of Model	Comment	Reference
Forward Reaction Models		
EQ3NR	Speciation-solubility part of EQ3/6. Extensive database with temperature range 0 to 300°C	Wolery (1983; 1992); Wolery and Daveler (1992)
GEOCHEM-PC	Designed for applications to soil and agricultural systems.	Sposito and Mattigod (1980), Sposito (1985)
MINEQL	An early water speciation code	Westall et al. (1976)
MINTEQA2	Large database of metallic species	Allison et al. (1991)
PHREEQE	Speciation calculation program calculates equilibrium among multiple phases	Parkhurst et al. (1980)
PHREEQC	C language version of PHREEQE with a number of improvements.	Parkhurst (1995)
PHRQPITZ	Combines PHREEQE and Pitzer equations for calculations of geochemical reactions in brines	Plummer et al. (1988)
SOILCHEM	Database contains mineral and organic species useful for soil systems	Sposito and Coves (1988)
SOLMINEQ.88	Database includes high temperature and pressure data	Kharaka et al. (1988)
WATEQ4F	Speciation calculation program with large mineral database	Ball and Nordstrom (1991)
Inverse Reaction Models		
BALANCE	Program for interpreting net geochemical mass-balance reactions between initial and final waters along a hydrologic flowpath. Predates NETPATH.	Parkhurst et al. (1982)
EQ6	Reaction path modeling part of EQ3/6. Uses titration processes (including fluid mixing), irreversible reaction in closed systems, irreversible reaction in some simple kinds of open systems, and heating or cooling processes; solves "single-point" thermo-dynamic equilibrium problems.	Wolery (1983, 1992); Wolery and Daveler (1992)
NETPATH	Interactive program for interpreting net geochemical mass-balance reactions between initial and final waters along a hydrologic flowpath.	Plummer et al. (1992, 1994)
Reaction-Transport Models		
ADNEUT	Acid-drainage neutralization computer program for simulating acid-drainage transport and neutralization.	Morin and Cherry (1988)
CHMTRNS	Mixed-differential algebraic model.	Noorishad et al. (1987)
FEED-FORWARD Method	Applicable in the presence of networks with any number of homogeneous and/or hetero-geneous, classical reaction segments that consist of three, at most binary participants.	Rubin (1990)
HYDROGEOCHEM	Sequential-iteration model.	Yeh and Tripathi (1992)
MST1D	Sequential-iteration model.	Engesgaard and Kipp (1992)
PHREEQM	One-dimensional mixing cell model simulates hydrodynamic dispersion.	Appelo and Willemsen (1987)
SATRA-CHEM	Direct-substitution model.	Lewis et al. (1987)

occur to a significant or insignificant extent along a designated flowpath. However, one code (EQ3/EQ6) incorporates reaction rate constants.

Another comparison may be made in which forward modeling tests the validity of suspected reactions based on thermodynamic considerations, whereas inverse modeling tests their feasibility based on mass balance considerations. Inverse modeling is especially useful after bench-scale column tests reveal the levels of various water quality parameters in the influent and effluent. Additionally, inverse modeling could be used to evaluate performance monitoring data after installation of the reactive cell.

Reaction-transport modeling is distinct, in that it can be used to simulate real transport processes, such as advection and dispersion, in addition to predicting groundwater chemistry. Thus, reaction-transport models may be especially useful for predicting the flowpath of both conservative and nonconservative species.

The availability of data is also a consideration in selecting a geochemical modeling code. Generally, fewer data are needed in forward modeling, whereas fairly complete data are required to achieve definitive results by inverse modeling. As with forward modeling, reaction-transport models may be run with limited chemical data; in addition, the hydraulic properties of the flow system must be understood.

C.2 AN ILLUSTRATION OF GEOCHEMICAL MODELING APPLIED TO REACTIVE BARRIERS EVALUATION

The following example illustrates the use of geochemical modeling in relation to predicting changes that might take place in groundwater chemistry that could affect the performance of a permeable barrier. The forward modeling code PHREEQC was used to demonstrate input criteria and possible output. Suppose that water samples are collected from an aquifer in which a permeable barrier is to be emplaced and are found to have the chemical makeup presented in Table C-2. The analysis indicated as "SOLUTION 1" in Table C-2 represents groundwater flowing through an aquifer containing the minerals dolomite [$MgCa(CO_3)_2$] and calcite ($CaCO_3$). The TDS content of this water (from an aquifer in the Knox dolomite of Tennessee) is 190 mg/L. Concentrations of the chemical constituents are indicated, including dissolved oxygen. The high redox potential is typical of an aerobic environment. In this example, equilibrium between groundwater, dolomite, and calcite is forced by the statements contained in the first "EQUILIBRIUM_PHASES" segment.

Results of speciation calculations and mineral saturation calculations are given in Tables C-3 and C-4, respectively.

Table C-3 shows that the chemistry of a natural water is quite complex when all of the common oxidation states of the elements are considered and the hydrolysis species, ion pairs, and complexes are evaluated. Based on these calculations, the program determines which minerals are likely to be at or near equilibrium with

TABLE C-2. Example of PHREEQC input

TITLE	Groundwater from Dolomitic Aquifer			
SOLUTION 1 Groundwater	units	mg/L		
	pH		7.4	
	pe		13.1	
	redox	$O(0)/O(-2)$		
	density		1.0	
	temp		15.0	
	O(0)		4.0	
	Mg		22.0	
	Na		0.4	
	K		1.2	
	Ca		40.0	
	Cl		2.0	
	S(6)	4.9 as SO_4		
	N(5)	4.8 as NO_3		
	Alkalinity	213.0 as HCO_3		
EQUILIBRIUM_PHASES				
	Calcite		0.0	1.0
	Dolomite		0.0	1.0
SAVE Solution 1				
END				
USE Solution 1				
EQUILIBRIUM_PHASES				
	FeMetal		0.0	1.0
END				

the groundwater. This is done by calculating the saturation index (SI) as the logarithm of the ion activity product (IAP) divided by the reaction equilibrium constant (K_T):

$$SI = \log (IAP/K_T) = \log (IAP) - \log (K_T)$$

Negative values of SI indicate undersaturation with respect to a particular solid phase; positive values of SI indicate oversaturation; and a value of zero indicates equilibrium. In practice, a solid may be in equilibrium when SI is within ±0.1 to 0.2 unit. Table C-4 indicates that the groundwater in this example is undersaturated with respect to all solid phases in the database other than calcite and dolomite, as stipulated in the input data. Aragonite can be ignored because it is unstable with respect to calcite under the conditions. The calculated concentrations of gaseous species are as follows:

$$\begin{aligned}
&CO_2(g) && 10^{-2.33} \text{ atm} = 0.0047 \text{ atm} \\
&O_2(g) && 10^{-0.99} \text{ atm} = 0.10 \text{ atm} \\
&\text{Other gases} && \text{negligible}
\end{aligned}$$

TABLE C-3. Speciation output for Solution 1 in Table C-2

Species	Molality	Activity	Species	Molality	Activity
OH^-	2.04E-07	1.88E-07	pH = 7.62		
H^+	2.58E-08	2.40E-08	pe = 13.8		
H_2O	5.55E+01	1.00E+00			
$C(-4)$	0.00E+00		$N(-3)$	0.00E+00	
CH_4	0.00E+00	0.00E+00	NH_4^+	0.00E+00	0.00E+00
$C(4)$	3.94E-03		NH_3	0.00E+00	0.00E+00
HCO_3^-	3.66E-03	3.38E-03	$NH_4SO_4^-$	0.00E+00	0.00E+00
CO_2	2.12E-04	2.12E-04	$N(0)$	5.88E-23	
$MgHCO_3^+$	3.00E-05	2.77E-05	N_2	2.94E-23	2.94E-23
$CaHCO_3^+$	2.73E-05	2.52E-05	$N(3)$	5.87E-18	
CO_3^{-2}	7.20E-06	5.26E-06	NO_2^-	5.87E-18	5.41E-18
$CaCO^3$	5.30E-06	5.31E-06	$N(5)$	7.74E-05	
$MgCO^3$	3.11E-06	3.11E-06	NO_3^-	7.74E-05	7.13E-05
$NaHCO^3$	3.04E-08	3.05E-08	Na	1.74E-05	
$NaCO^{3-}$	1.01E-09	9.31E-10	Na^+	1.74E-05	1.60E-05
Ca	1.01E-03		$NaHCO_3$	3.04E-08	3.05E-08
Ca^{+2}	9.68E-04	7.06E-04	$NaSO_4^-$	2.55E-09	2.35E-09
$CaHCO_3^+$	2.73E-05	2.52E-05	$NaCO_3^-$	1.01E-09	9.31E-10
$CaCO_3$	5.30E-06	5.31E-06	$NaOH$	4.41E-12	4.42E-12
$CaSO_4$	4.00E-06	4.00E-06	$O(0)$	2.50E-04	
$CaOH^+$	5.30E-09	4.89E-09	O_2	1.25E-04	1.25E-04
Cl	5.64E-05		$S(-2)$	0.00E+00	
Cl^-	5.64E-05	5.20E-05	HS^-	0.00E+00	0.00E+00
$H(0)$	0.00E+00		H_2S	0.00E+00	0.00E+00
H_2	0.00E+00	0.00E+00	S^{-2}	0.00E+00	0.00E+00
K	3.07E-05		$S(6)$	5.10E-05	
K^+	3.07E-05	2.83E-05	SO_4^{-2}	4.30E-05	3.13E-05
KSO_4^-	5.95E-09	5.49E-09	$MgSO_4$	4.06E-06	4.06E-06
KOH	4.09E-12	4.09E-12	$CaSO_4$	4.00E-06	4.00E-06
Mg	1.03E-03		KSO_4^-	5.95E-09	5.49E-09
Mg^{+2}	9.88E-04	7.24E-04	$NaSO_4^-$	2.55E-09	2.35E-09
$MgHCO_3^+$	3.00E-05	2.77E-05	HSO_4^-	6.41E-11	5.91E-11
$MgSO_4$	4.06E-06	4.06E-06	$NH_4SO_4^-$	0.00E+00	0.00E+00
$MgCO_3$	3.11E-06	3.11E-06			
$MgOH^+$	4.67E-08	4.31E-08			

    The next part of the program simulates the changes that would take place in the groundwater if it were to come into contact with zero-valent iron. The system is closed to the atmosphere during this step. After saving the calculations shown in Table C-3, the computer code applies the constraint identified under "USE Solution 1" in Table C-2, where FeMetal implies the following reaction:

$$Fe \rightleftharpoons Fe^{+2} + 2e^-$$

    The WATEQ4F database (Ball et al., 1987) was used to obtain values for log K (15.114) and ΔH (-21.300 kcal/mol); the standard enthalpy of reaction, ΔH, is needed to calculate log K at temperatures other than at 25°C, by the Van't Hoff

TABLE C-4. Mineral saturation indices associated with Table C-3

Phase	SI	log IAP	log K$_T$	Formula
Anhydrite	−3.32	−7.66	−4.34	CaSO$_4$
Aragonite	−0.15	−8.43	−8.28	CaCO$_3$
Calcite	0.00	−8.43	−8.43	CaCO$_3$
CH$_4$(g)	−146.34	−191.73	−45.4	CH$_4$
CO$_2$(g)	−2.33	−20.52	−18.19	CO$_2$
Dolomite	0.00	−16.85	−16.85	CaMg(CO$_3$)$_2$
Gypsum	−3.07	−7.66	−4.58	CaSO$_4$:2H$_2$O
H$_2$(g)	−42.84	−42.8	0.03	H$_2$
H$_2$S(g)	−147.77	−190.96	−43.19	H$_2$S
N$_2$(g)	−19.31	−237.55	−218.25	N$_2$
NH$_3$(g)	−70.69	−182.98	−112.29	NH$_3$
O$_2$(g)	−0.99	85.61	86.6	O$_2$
Sulfur	−110.97	−148.16	−37.18	S

equation. Table C-5 lists concentrations of dissolved species in equilibrium with Fe^0 and Table C-6 gives mineral saturation data. Table C-5 contains a greater number of species than does Table C-3 because Fe is a component of the system and additional complexes are formed with the +2 and +3 valence states of iron.

Tables C-5 and C-6 show that several significant changes in mineral saturation have occurred. In this step Fe is the only solid phase required to be in equilibrium with the groundwater. As a result, pH has increased from 7.62 to 11.92, pe has become negative, and dissolved oxygen is depleted. Calculations of aqueous species show that redox-sensitive species such as carbon, hydrogen, nitrogen, and sulfur are reduced. The reduction products are methane, hydrogen gas, ammonia and ammonium ion, and aqueous sulfide species. Mineral saturation indices have changed as well. The solution now is slightly undersaturated with respect to calcite and quite undersaturated with respect to dolomite. However, the groundwater is well oversaturated with respect to brucite (magnesium hydroxide), amorphous ferric hydroxide, goethite, ferrous sulfide precipitate, and siderite (ferrous carbonate). Based on the calculations by PHREEQC, compounds that are oversaturated (SI > 0) are likely to precipitate out of solution in the reactive cell. Some other compounds such as hematite, mackinawite, and pyrite are not likely to precipitate even though they are oversaturated, due to slow reaction kinetics. Similarly, sulfate often does not become reduced in the presence of iron unless conditions are conducive to a biological mechanism of reduction.

The calculations described in this example provide some insight into the types of reactions that should be considered when evaluating performance and longevity of a reactive cell. Forward geochemical models, such as PHREEQC, can predict the behavior of a system based on an assumption of equilibrium. Therefore, the output does not guarantee (1) whether the reactions will actually take place and, if they do, (2) to what extent they will occur. Nevertheless, forward modeling is a significant initial activity that should be performed to estimate the potential for side reactions that could impact the operation of a permeable barrier.

TABLE C-5. Speciation output for equilibrium of groundwater with Fe^0

Species	Molality	Activity	Species	Molality	Activity
OH^-	4.22E-03	3.77E-03	pH = 11.92		
H^+	1.32E-12	1.20E-12	pe = -9.97		
H_2O	5.55E+01	1.00E+00			
C(-4)	3.93E-03		**H(0)**	1.83E-07	
CH_4	3.93E-03	3.94E-03	H_2	9.13E-08	9.16E-08
C(4)	1.02E-05		**K**	3.07E-05	
CO_3^{-2}	3.96E-06	2.58E-06	K^+	3.06E-05	2.74E-05
$FeCO_3$	3.17E-06	3.18E-06	KOH	7.90E-08	7.93E-08
$CaCO_3$	2.19E-06	2.19E-06	$KSO4^-$	7.41E-17	6.64E-17
$MgCO_3$	7.56E-07	7.58E-07	**Mg**	1.03E-03	
HCO_3^-	9.21E-08	8.28E-08	Mg^{+2}	5.47E-04	3.59E-04
$CaHCO_3^+$	5.78E-10	5.19E-10	$MgOH^+$	4.78E-04	4.28E-04
$NaCO_3^-$	4.95E-10	4.43E-10	$MgCO_3$	7.56E-07	7.58E-07
$FeHCO_3^+$	4.75E-10	4.25E-10	$MgHCO_3^+$	3.76E-10	3.36E-10
$MgHCO_3^+$	3.76E-10	3.36E-10	$MgSO_4$	2.51E-14	2.52E-14
$NaHCO_3$	7.21E-13	7.23E-13	**N(-3)**	7.75E-05	
CO_2	2.59E-13	2.60E-13	NH_3	7.71E-05	7.73E-05
Ca	1.01E-03		NH_4^+	3.86E-07	3.43E-07
Ca^{+2}	9.11E-04	5.94E-04	$NH_4SO_4^-$	1.93E-18	1.73E-18
$CaOH^+$	9.20E-05	8.23E-05	**N(0)**	1.83E-10	
$CaCO_3$	2.19E-06	2.19E-06	N_2	9.13E-11	9.16E-11
$CaHCO_3^+$	5.78E-10	5.19E-10	**N(3)**	0.00E+00	
$CaSO_4$	4.19E-14	4.20E-14	NO_2^-	0.00E+00	0.00E+00
Cl	5.65E-05		**N(5)**	0.00E+00	
Cl^-	5.65E-05	5.04E-05	NO_3^-	0.00E+00	0.00E+00
$FeCl^+$	3.99E-09	3.58E-09	**Na**	1.74E-05	
$FeCl^{+2}$	5.06E-31	3.25E-31	Na^+	1.73E-05	1.55E-05
$FeCl_2^+$	1.14E-34	1.02E-34	$NaOH$	8.55E-08	8.57E-08
$FeCl_3$	5.11E-40	5.13E-40	$NaCO_3^-$	4.95E-10	4.43E-10
Fe(2)	7.09E-03		$NaHCO_3$	7.21E-13	7.23E-13
$FeOH^+$	7.00E-03	6.20E-03	$NaSO_4^-$	3.18E-17	2.85E-17
Fe^{+2}	7.82E-05	5.14E-05	**O(0)**	0.00E+00	
$Fe(HS)_2$	1.47E-05	1.48E-05	O_2	0.00E+00	0.00E+00
$FeCO_3$	3.17E-06	3.18E-06	**S(-2)**	5.11E-05	
$Fe(HS)_3^-$	3.23E-08	2.89E-08	HS^-	2.01E-05	1.80E-05
$FeCl^+$	3.99E-09	3.58E-09	$Fe(HS)_2$	1.47E-05	1.48E-05
$FeHCO_3^+$	4.75E-10	4.25E-10	S^{-2}	1.38E-06	8.93E-07
$FeSO_4$	2.95E-15	2.95E-15	$Fe(HS)_3^-$	3.23E-08	2.89E-08
$FeHSO_4^+$	2.55E-26	2.28E-26	H_2S	2.88E-10	2.89E-10
Fe(3)	6.26E-03		**S(6)**	6.73E-13	
$Fe(OH)_4^-$	6.25E-03	5.60E-03	SO_4^{-2}	6.03E-13	3.91E-13
$Fe(OH)_3$	1.11E-05	1.11E-05	$CaSO_4$	4.19E-14	4.20E-14
$Fe(OH)_2^+$	1.82E-10	1.62E-10	$MgSO_4$	2.51E-14	2.52E-14
$FeOH^{+2}$	1.35E-18	8.69E-19	$FeSO_4$	2.95E-15	2.95E-15
Fe^{+3}	7.06E-28	2.96E-28	KSO_4^-	7.41E-17	6.64E-17
$FeCl^{+2}$	5.06E-31	3.25E-31	$NaSO_4^-$	3.18E-17	2.85E-17
$Fe_2(OH)_2^{+4}$	1.84E-34	3.12E-35	$NH_4SO_4^-$	1.93E-18	1.73E-18
$FeCl_2^+$	1.14E-34	1.02E-34	HSO_4^-	4.12E-23	3.69E-23
$FeSO_4^+$	1.13E-36	1.01E-36	$FeHSO_4^+$	2.55E-26	2.28E-26
$FeCl_3$	5.11E-40	5.13E-40	$FeSO_4^+$	1.13E-36	1.01E-36
$Fe_3(OH)_4^{+5}$	0.00E+00	0.00E+00	$Fe(SO_4)_2^{2-}$	0.00E+00	0.00E+00
$Fe(SO_4)_2^-$	0.00E+00	0.00E+00	$FeHSO_4^{+2}$	0.00E+00	0.00E+00
$FeHSO_4^{+2}$	0.00E+00	0.00E+00			

TABLE C-6. Mineral saturation indices associated with Table C-4

Phase	SI	log IAP	log K$_T$	Formula
Anhydrite	−11.3	−15.63	−4.34	CaSO$_4$
Aragonite	−0.54	−8.81	−8.28	CaCO$_3$
Brucite	2.87	20.4	17.53	Mg(OH)$_2$
Calcite	−0.38	−8.81	−8.43	CaCO$_3$
CH$_4$[a]	0.37	−45.03	−45.4	CH$_4$
CO$_2$[a]	−11.24	−29.43	−18.19	CO$_2$
Dolomite	−1.00	−17.85	−16.85	CaMg(CO$_3$)$_2$
Fe(OH)$_3$[b]	3.35	21.5	18.16	Fe(OH)$_3$
FeMetal	0.00	15.66	15.66	Fe
FeS (ppt)[c]	6.8	−32.29	−39.09	FeS
Goethite	9.24	21.5	12.27	FeOOH
Gypsum	−11.05	−15.63	−4.58	CaSO$_4$:2H$_2$O
H$_2$[a]	−3.93	−3.9	0.03	H$_2$
H$_2$S[a]	−8.66	−51.85	−43.19	H$_2$S
Hematite	19.7	43.01	23.31	Fe$_2$O$_3$
Jarosite-K	−32.02	−0.63	31.38	KFe$_3$(SO$_4$)$_2$(OH)$_6$
Mackinawite	7.54	−32.29	−39.83	FeS
Melanterite	−14.36	−16.7	−2.34	FeSO$_4$:7H$_2$O
N$_2$[a]	−6.81	−10.04	−3.23	N$_2$
NH$_3$[a]	−6.09	−10.87	−4.78	NH$_3$
O$_2$[a]	−78.8	7.8	86.6	O$_2$
Portlandite	−2.97	20.62	23.59	Ca(OH)$_2$
Pyrite	8.88	−80.24	−89.13	FeS$_2$
Siderite	0.95	−9.88	−10.83	FeCO$_3$
Sulfur	−10.76	−47.95	−37.18	S

(a) Gas.

(b) Amorphous solid.

(c) ppt = parts per trillion.

C.3 APPENDIX C REFERENCES

Allison, J.D., D.S. Brown, and K.J. Novo-Gradac. 1991. *MINTEQA2/PRODEFA2, A Geochemical Assessment Model for Environmental Systems: Version 3.0 User's Manual.* EPA/600/3-91/021. U.S. Environmental Protection Agency, Office of Research and Development, Washington, DC.

Appelo, C.A.J., and A. Willemsen. 1987. "Geochemical Calculations and Observations on Salt Water Intrusions. I. A Combined Geochemical/Mixing Cell Model." *J. Hydrol. 94*: 313-330.

Ball, J.W., and D.K. Nordstrom. 1991. *User's Manual for WATEQ4F, with Revised Thermodynamic Data Base and Test Cases for Calculating Speciation of Major, Trace, and Redox Elements in Natural Waters.* U.S. Geological Survey Open-File Report 91-183.

Ball, J.W., D.K. Nordstrom, and D.W. Zachmann. 1987. *WATEQ4F—A Personal Computer FORTRAN Translation of the Geochemical Model WATEQ2 with Revised Database.* U.S. Geological Survey Open File Report 87-50.

Engesgaard, P., and K.L. Kipp. 1992. "A Geochemical Transport Model for Redox-Controlled Movement of Mineral Fronts in Groundwater Flow Systems: A Case of Nitrate Removal by Oxidation of Pyrite." *Water Resources Res. 28* (10): 2829-2843.

Kharaka, Y.K., W.D. Gunter, P.K. Aggarwal, W.H. Perkins, and J.D. DeBraal. 1988. *SOLMINEQ.88: A Computer Program for Geochemical Modelling of Water-Rock Reactions.* U.S. Geological Survey Water-Resources Investigations Report 88-4227, Menlo Park, CA.

Lewis, F.M., C.I. Voss, and J. Rubin. 1987. "Solute Transport with Equilibrium Aqueous Complexation and Either Sorption or Ion Exchange: Simulation Methodology and Applications." *Journal of Hydrology 90*: 81-115.

Mangold, D.C., and C-F. Tsang. 1991. "A Summary of Subsurface Hydrological and Hydrochemical Models." *Reviews of Geophysics 29*: 51-79.

Morin, K.A., and J.A. Cherry. 1988. "Migration of Acidic Groundwater Seepage From Uranium-Tailings Impoundments. 3: Simulation of the Conceptual Model with Application to Seepage Area A." *Journal of Contaminant Hydrology 2*: 323-342.

Noorishad, J., C.L. Carnahan, and L.V. Benson. 1987. *A Report on the Development of the Non-Equilibrium Reactive Transport Code CHMTRNS.* Report LBL-22361, Lawrence Berkeley Laboratory, Berkeley, CA.

Parkhurst, D.L. 1995. *User's Guide to PHREEQC—A Computer Program for Speciation, Reaction-Path, Advective-Transport, and Inverse Geochemical Calculations.* USGS 95-4227, Lakewood, CO.

Parkhurst, D.L., L.N. Plummer, and D.C. Thorstenson. 1982. *BALANCE—A Computer Program for Calculating Mass Transfer for Geochemical Reactions in Groundwater.* Water-Resources Investigation 82-14. U.S. Geological Survey.

Parkhurst, D.L., D.C. Thorstensen, and L.N. Plummer. 1980. *PHREEQE—A Computer Program for Geochemical Calculation.* Water Resources Investigation 80-96, U.S. Geological Survey.

Plummer, L.N., D.L. Parkhurst, G.W. Fleming, and S.A. Dunkle. 1988. Water-Resources Investigations Report 88-4153. U.S. Geological Survey.

Plummer, L.N., E.C. Prestemon, and D.L. Parkhurst. 1992. "NETPATH: An Interactive Code for Interpreting NET Geochemical Reactions from Chemical and Isotopic Data Along a Flow PATH." In Y. Kharaka and A.S. Maest (Eds.), *Proceedings, 7th International Symposium on Water-Rock Interaction,* pp. 239-242. Park City, UT, July 9-23, Balkema, Rotterdam.

Plummer, L.N., E.C. Prestemon, and D.L. Parkhurst. 1994. *An Interactive Code (NETPATH) for Modeling NET Geochemical Reactions Along a Flow PATH— Version 2.0.* Water-Resources Investigations Report 94-4169. U.S. Geological Survey.

Rubin, J. 1990. "Solute Transport with Multisegment, Equilibrium-Controlled, Reactions: A Feed-Forward Simulation Method." *Water Resour. Res. 26*: 2029-2055.

Sposito, G. 1985. "Chemical Models of Inorganic Pollutants in Soils." *Critical Reviews in Environmental Control 15*(1): 1-24.

Sposito, G., and J. Coves. 1988. *SOILCHEM: A Computer Program for the Calculation of Chemical Speciation in Soils.* Report: Kearney Foundation, University of California, Riverside and Berkeley, CA.

Sposito, G., and S.V. Mattigod. 1980. Report, Kearney Foundation of Soil Science. University of California.

Westall, J., J.L. Zachary, and F.M.M. Morell. 1976. Massachusetts Institute of Technology, Department of Civil Engineering Technical Note 18.

Wolery, T.J. 1983. *EQ3NR—A Computer Program for Geochemical Aqueous Speciation-Solubility Calculations: User's Guide and Documentation.* Lawrence Livermore Laboratory Report UCRL-53414.

Wolery, T.J. 1992. *EQ3NR, A Computer Program for Geochemical Aqueous Speciation-Solubility Calculations: Theoretical Manual, User's Guide and Related Documentation (Version 7.0).* Report UCRL-MA-110662 PT III, Lawrence Livermore National Laboratory, CA.

Wolery, T.J., and S.A. Daveler. 1992. *EQ6, a Computer Program for Reaction Path Modeling of Aqueous Geochemical Systems: Theoretical Manual, User's Guide, and Related Documentation (Version 7.0). Part 4.* Lawrence Livermore National Laboratory, Livermore, CA. Report No.: UCRL-MA-110662-PT.4. October 9.

Yeh, G.T., and V.S. Tripathi. 1992. "A Model for Simulating Transport of Reactive Multispecies Components: Model Development and Demonstration." *Water Resources Res. 27*(12): 3075-3094.

CONSTRUCTION QUALITY CONTROL

The effectiveness and long-term performance of either a permeable or impermeable barrier depends on the level of construction quality control (CQC) that is implemented. This section will address the more commonly used barrier technologies in environmental applications such as slurry walls, deep soil mixing, and jetting. In addition, CQC issues for sealable-joint sheet piles will be addressed because this technique is becoming more widely used.

D.1 SLURRY WALLS

Steps that should be taken prior to construction include the development of a CQC plan that reflects the technical aspects of the barrier design. After the depth of the wall is specified, the plan should be geared towards the two important aspects of the wall design, conforming impermeability and wall continuity. This is carried out through a series of CQC tests conducted prior to wall construction to accurately provide the specifications of the slurry makeup and the backfill mix, specifically, its compatibility with contaminants, its reactions to temperature, and its ability to achieve the desired permeability. It should address remedial techniques for substandard construction practices and for any other problems that might arise during implementation.

Quality control (QC) considerations and guidelines for slurry walls can be found in Bell and Sisley (1992).

D.2 SEALABLE-JOINT SHEET PILES

CQC issues for sealable-joint sheet piles are reported in the "Waterloo Barrier™ Pile Driving and Joint Sealing General Specifications" provided by Jowett (1996). These include:

- Pile Driving Plan. This plan outlines detailed pile placement; splicing requirements and details; method to achieve verticality within 1%; QC measures; joint preparation prior to sealing; and grout materials, mixing, and placement.

- Mill test documentation for piling to be used on project.

- Manufacturer's data that indicate the structural properties of piling sections(s) to be used.

- Proposed welding procedures and certification of welders.

- Horizontal alignment and plumbness tolerances.

- If the contractor chooses to drive sheet piles in doubles, a cone shall be employed to prevent soil from entering the mated (center) joint. The contractor will be responsible for the fabrication and installation of the cone for each paired sheet pile set. Specific dimensions of the cones will be based on actual rolled sheets. Foot plates will be welded to the base of the female joint of the paired set.

- Sheet pile handling procedures for field installation.

- Location and tolerances.

- Sheet pile setup.

- Joint sealing procedures.

- Provide accurate records of each sheet pile installed. Submitted records will include the following information:

    — Pile identification number
    — Date and time of driving
    — Model of hammer and energy rating
    — Elevation at top of pile
    — Length of sheet pile in the ground when driving is complete
    — Rate of penetration in ft/minute
    — Detailed remarks concerning alignment, obstructions, etc.
    — Plumbness record of each sheet pile installed
    — Joint flushing record for each joint installed.

D.3 APPENDIX D REFERENCES

Bell, R.A., and J.L. Sisley. 1992. "Quality Control of Slurry Cutoff Wall Installations." In D.B. Paul, R.R. Davidson, and N.J. Cavalli (Eds.), *Slurry Walls: Design, Construction and Quality Control, ASTM STP 1129.* American Society for Testing and Materials, Philadelphia, PA.

Jowett, R. 1996. Personal communication and marketing literature from R. Jowett of Waterloo Barrier, Inc., Ontario, Canada.

INDEX